AFTER THE
ICE AGE

AFTER THE
ICE AGE

The Return of Life to
Glaciated North America

E. C. PIELOU

The University of Chicago Press ◆ Chicago and London

The University of Chicago Press, Chicago 60637
The University of Chicago Press, Ltd., London
© 1991 by The University of Chicago
All rights reserved. Published 1991
Paperback edition 1992
Printed in the United States of America
99 98 97 96 95 94 93 92   5 4 3

Library of Congress Cataloging-in-Publication Data

Pielou, E. C.
    After the Ice Age : the return of life to glaciated North America / E. C. Pielou.
        p. cm.
    Includes bibliographical references and index.
    ISBN 0-226-66811-8 (cloth)   ISBN 0-226-66812-6 (paperback)
    1. Paleobiogeography—North America.   2. Paleontology—Holocene.   3. Glacial
epoch—North America.   4. Paleoecology—North America.   I. Title.
QE721.2.P24P54   1991
560'.1'78—dc20   90-11024
                                                                         CIP

FOR PATRICK

Contents

PART FIVE

OUR PRESENT EPOCH, THE HOLOCENE

Prologue

With the approach of the year 2000 A.D. and the end of the twentieth century, as the western world measures time, speculation mounts as to what the future holds. No doubt our forebears indulged in similar speculations, as each succeeding century drew to a close for many centuries back into the past. But now, as never before, people are speculating on the future of the nonhuman world as well as the human. The development of human history has always been governed by the setting—the natural environment—in which it has taken place. In the past this setting changed so slowly that it could be regarded as static. Predictions about humanity's future did not need to take account of changing climate, spreading deserts, rising sea levels, disappearing forests, and the like. Now, as we are all aware, changes of these kinds are likely to affect our future profoundly. We are now well into the population explosion that has threatened us at least since the time Thomas Robert Malthus (1766–1834) first published his dire warnings, and the exploding population is quickly degrading its own environment.

But environmental change, albeit at a slower rate, is not new, even though history books seldom mention it. In a large part of the world, recorded human history has been too short to encompass much change. If we look at a longer time-span, however, from the time when the last ice age was at its height down to the present day, the tremendous magnitude of the changes our world has undergone becomes apparent.

This transformation has been greatest of all in northern North America, where the ice sheets of the last ice age were far larger (by fifty percent, at least) than those in Europe and Asia combined; and where, at its maximum, a little less than 20,000 years ago, the ice

1

formed an unbroken expanse about as large as Antarctica. It covered nearly all of Canada and the northern tier of the United States.

This book is concerned with what happened in this land between then and now, with the way the environment changed as the ice melted, and with the way life returned to what had been a lifeless semicontinent. Most people are aware that half of North America was once covered by an ice sheet kilometers thick, and they marvel at the contrast between conditions then and now. Not nearly so many try to visualize the remarkable scenery of intervening times while the ice was in the process of melting. Sometimes windswept deserts developed when fierce winds, unimpeded by vegetation, picked up the sand and dust uncovered by the retreating ice sheets and built it into huge dunes. At other times forests invaded the margin of the ice itself. For long periods the Midwest was covered by lakes greater by far than the modern Great Lakes, on which massive icebergs floated and whose northern shores consisted of tall cliffs of ice. Swift rivers flowed along the ice margin and shifted their courses abruptly as the ice melted. Sometimes dams of ice collapsed, allowing tremendous floods to cascade across the land, shifting huge boulders with their torrential flow. Melting ice made the land "active" in a way that has no parallel in modern times.

As the ice disappeared, life found its way into this strange landscape. Vegetation gradually took over the newly exposed ground, enabling animal immigrants to survive. The animals, especially the large mammals, would have been as unfamiliar as the scenery to modern eyes. Mastodons, mammoths, sabertooth cats, and giant short-faced bears were among the most spectacular. The small human population of 10,000 years ago *may* be to blame, at least in part, for their extinction. Our continent is the poorer for their disappearance; a knowledge of what we have lost should spur us to greater efforts to protect what remains; the grizzly bears, polar bears, cougars, bighorns, and other species that still survive, precariously, in what wild land remains to us.

A history of the last 20,000 years in northern North America, ending 200 to 300 years ago when the population explosion began, is indeed a history of wild land. It is also a history of an astonishing transformation, as the following chapters will show.*

---

*Animals and plants are given their English names in the text. The scientific (Latin) equivalents are in the appendixes.

PART ONE

PRELIMINARIES

PART ONE

PRELIMINARIES

1

The Physical Setting

Every story has a context. The context of the story in this book is the land, the fresh water, the surrounding sea, the ice, and the climate of northern North America during the past 20,000 years. This is the period since the ice sheets of the most recent ice age were at their greatest extent. It is the period during which the ice in North America has almost completely melted. Almost, but not quite. Some still remains, in the form of glaciers and ice fields in the western mountains and some ice caps in the Arctic islands, as shown in fig. 1.1.

So much for the immediate context, but the context itself has a context. Twenty thousand years is only a short period in geological terms; it is worth considering how climatic change during the period covered in this book fits into the pattern of change over a sizable fraction of the lifetime of the earth, specifically over the past one billion (1,000,000,000) years, which is probably rather more than one-fifth of the earth's lifetime to date.

Research into the climates of the past shows that climatic change is, not surprisingly, never ending. It happens on several scales: there are large variations, taking many millions of years; lesser variations within the large ones, taking tens or hundreds of thousands of years; and so on, with a succession of variations of decreasing magnitude and duration forming a nested series. In this section we consider only the two largest scales.

In the past billion years there have been a number of cold periods, popularly called ice ages, but the term *ice age* is an imprecise, general term. By itself it can mean a long cool period, of ten million years' duration, or an extra cold period, of a mere 100,000 years or so, within one of the long cool periods. To keep the distinction clear, it helps to use separate names for ice ages of different rank. Those of highest rank are called *glacial ages*, whereas those of second rank are called *glaciations*. As an introduction to all that follows, here is a very brief outline of what is now known or surmised about the timing, and the causes, of glacial ages and glaciations.[1]

To make the relative lengths of enormous stretches of time easy to visualize, let us use as a model one decade to represent the past billion years. During this time there have been two complete glacial ages and the beginnings of a third. On the scale of the model, a glacial age lasts a month or two; one occurred eight or nine years ago; a second, between two and three years ago; the third (which we live in) began only six or seven days ago; we are still near the beginning of it. Still on the scale of the model, the climate has been comparatively warm for nine years (at least) when there was not a glacial age in progress.

A glacial age is not continuously cold. This is known with cer-

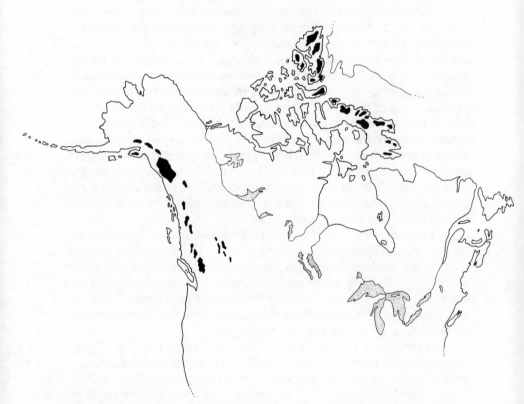

FIGURE 1.1: The present extent of perennial ice in northern North America. Black areas are ice-covered; stippled areas are freshwater lakes.

tainty for the current glacial age and is probably true of the earlier ones. During the six or seven days (on the scale of the model) of the current glacial age that have elapsed so far, there have been about nineteen or twenty intensely cold periods (*glaciations*), each of about six hours' duration, separated by short, comparatively mild respites (*interglacials*) of about one and a half hours duration. We are at this moment living in an interglacial and are within minutes of the end of it.

Now for causes. Many theories purporting to account for climatic change have been proposed since the scientific study of earth history first began; by far the most persuasive of the modern theories is that glacial ages come and go because of the constantly shifting pattern of the earth's tectonic plates. The argument is as follows. For most of the earth's history the continents and oceans have been so arranged that warm ocean currents from the tropics were able to flow easily into the north and south polar regions; there were no barriers to the flow of currents from low to high latitudes. That is not how matters stand now. At present, a continent, Antarctica, covers a large area centered on the South Pole; the Arctic Ocean, centered on the North Pole, is almost cut off from surrounding oceans because of the way the northern continents are arrayed around it. This is why we are now experiencing a glacial age; warm ocean currents are prevented, by land barriers, from circulating the sun's heat away from tropical and temperate latitudes into the polar regions. As a result, summers are cool at high latitudes; they are not warm enough to melt all of the ice and snow that forms each winter. Therefore, the ice and snow accumulate, year after year, and we have a glacial age.

Presumably the same mechanism accounts for earlier glacial ages. The last complete glacial age before our present one (which is still in progress) happened about 250 million years ago (2.5 years on the model scale in which one billion years is represented by ten years). It does not follow that the pattern of continents and oceans was the same then as now, only that the continents were so placed as to block the free exchange of tropical and polar ocean waters.

Because a far greater proportion of the earth's surface is covered by sea than by land, patterns that produce a glacial age are comparatively uncommon, and the earth has experienced the conditions of a glacial age for only a small proportion of its total history. The duration of a glacial age, when one happens, is governed by the speed at which the tectonic plates drift; probably each one lasts for ten or more million years. We are only about two million years into the current one.

Now we come to glaciations. Why should the climate vary within a glacial age, so that for parts of the time enormous ice sheets cover lands in temperate latitudes (that is, there are "ice ages"), whereas at other times, during the interglacials, the big ice sheets shrink markedly? They do not necessarily disappear entirely; the ice sheets of Antarctica and Greenland have survived thus far in our present interglacial (which is probably reaching its end) and will no doubt persist into the next glaciation.

It is believed that glaciations and interglacials alternate regularly. The climate has a 100,000-year cycle, in which glaciations lasting 60,000 to 90,000 years alternate with interglacials lasting between 40,000 and 10,000 years, respectively. (These values are rounded for convenience; now that we are considering the geologically recent past, durations and dates are given in "real" time; the model time scale used previously will be dropped.)

What must be explained is the reason for the 100,000-year climatic cycle. In light of recent research, it seems almost certain that the climatic cycle is forced by a 100,000-year cycle in the behavior of the earth in its annual orbit around the sun. This astronomical cycle is known as the Milankovitch cycle after its discoverer.[2] It is, itself, the resultant of three other cycles: a 105,000-year cycle in the shape of the earth's orbit, from a more elongated to a less elongated ellipse, and back; a 41,000-year cycle in the tilt of the earth's axis relative to an axis perpendicular to the orbital plane (the tilt is known as the obliquity of the ecliptic); and a 21,000-year cycle, called the precession of the equinoxes, during which the moment in the year at which the earth is closest to the sun, as it traverses its elliptical orbit, shifts forward from January through February, then March, and so on, around to January again.

The climatic effect of these cycles is a variation in the degree of contrast between summer and winter temperatures. There is no change in the total amount of solar radiation falling in one year on the whole earth; what is affected is the partitioning of this radiation between the northern and southern hemispheres and between high latitudes and low. At one extreme of a Milankovitch cycle there is a combination of comparatively warm summers and cold winters at high northern latitudes. At the other extreme there is a combination of cool summers and mild winters. The same sequence of events takes place at high southern latitudes as well. It is not yet known whether, and by how much, the glaciations in the two hemispheres are out of step with each other; this depends on the different climatic effects of the three cycles that together make up the Milankovitch cycle. In all that follows we concentrate on the climate of our own hemisphere, the northern.

The degree of contrast between the seasons is the climatic factor that, more than any other, accounts for the formation and disappearance of ice sheets over the land during a glacial age. This contrast depends, at all latitudes, on the eccentricity of the earth's orbit (which varies in the 105,000-year cycle) and on the season when the earth is closest to the sun (which varies in the 21,000-year precession cycle). The 41,000-year cycle in the obliquity (or tilt) of the earth's axis has most effect at high latitudes; the contrast between the seasons at high latitudes is much greater when the tilt is large (24.4 degrees) than when it is small (21.8 degrees), but the angle of tilt makes little difference to seasonal contrasts at low latitudes.

Throughout most of a glacial age (that is, during the glaciations), when the contrast between seasons is comparatively slight, summer temperatures are not high enough for the previous winter's snow and ice to melt. They accumulate year after year, inexorably building up huge continental ice sheets in temperate latitudes, even though the winters are relatively mild. But during one short segment of each Milankovitch cycle, the contrast between the seasons is great enough, that is to say the summers are hot enough, for each winter's snow and ice to melt before the onset of the succeeding winter (except in polar land masses like Antarctica and Greenland). It is the high summer temperatures that count; the winters are extra cold as well during this phase of the cycle, but that does not affect the completeness of the thaw each summer. The result is an interglacial such as the one now coming to an end.

The change from a glaciation to an interglacial is not abrupt, of course. As summers become imperceptibly warmer year after year, near one extreme of a Milankovitch cycle, the ice begins to melt. The current Milankovitch cycle reached this stage about 18,000 years ago and the results, for North America, have been dramatic. The series of maps in figures 1.2 through 1.5 and finally fig. 1.1 show how the two huge ice sheets of the most recent glaciation, the Laurentide and the Cordilleran, melted away.

As the ice sheets vanished, they left a growing expanse of bare ground where, at first, nothing lived. Over the millennia, the area has been colonized by an enormous number of plants and animals whose permanent home is in the low-latitude regions that are ice free during the glaciations. The invaders managed to occupy the newly exposed land and water as it became available, and now they form the biosphere of most of northern North America.

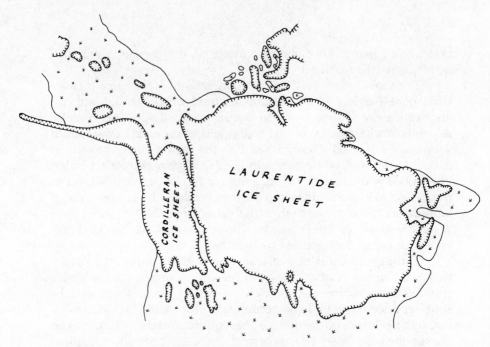

FIGURE 1.2: Northern North America 18,000 years ago.[3]

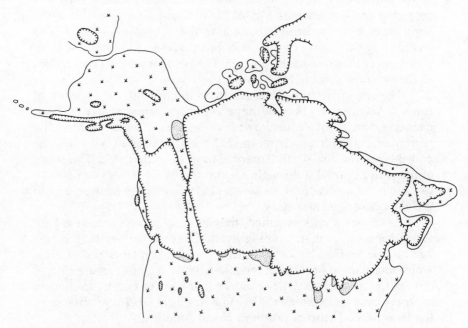

FIGURE 1.3: Northern North America 13,000 years ago.

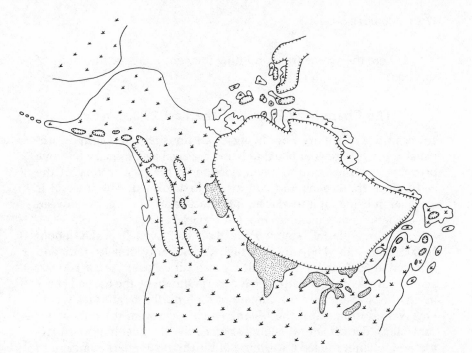

FIGURE 1.4: Northern North America 10,000 years ago.

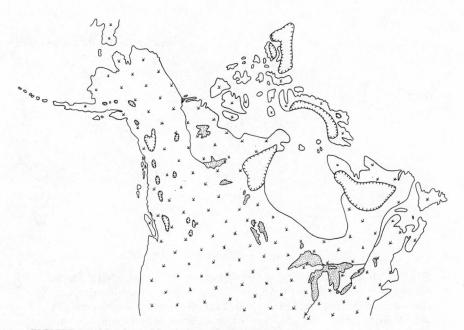

FIGURE 1.5: Northern North America 7,000 years ago.

Where they came from and how they got there are the topics of this book.

The Changing Climate of the Last 20,000 Years

To recapitulate: we are now living in an unusual interval in an unusual age. The age is unusual in being a glacial age; glacial ages have probably occupied only a small fraction of the past history of the earth since the oceans and continents first formed. The interval is unusual in being an interglacial; interglacials are short gaps between the glaciations that make up most of a glacial age.

Our interglacial is now more than half over. It peaked about 10,000 years ago. The way in which the quantity of solar radiation received at 65 degrees north latitude has varied over the past 20,000 years because of the Milankovitch cycle is shown by the dashed curve in fig. 1.6. Presumably the climate of high northern latitudes would have varied to match if the Milankovitch cycle were the only factor controlling the climate. It is almost certainly the most important factor, and when it caused a warming of northern summers sufficient to melt the previous winters' snows, the last glaciation quickly gave way to the current interglacial. But as the solid line in the figure

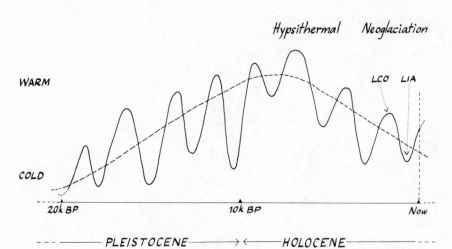

FIGURE 1.6: Climatic variations during the last 20,000 years.[4] The dashed curve shows how solar radiation reaching the earth at sixty-five degrees north latitude varied because of cyclical variations in the earth's orbit (the Milankovitch cycle). The solid line is a speculative reconstruction of actual temperature variation. LCO, Little Climatic Optimum; LIA, Little Ice Age.

shows, there were other, lesser climatic variations superimposed on the main one, and they caused the alternate warm and cool periods shown in the graph. In particular, there appears to have been a cyclical variation with a period of about 2,500 years throughout the last 10,000 years; it may have been caused by cyclical variations in the sun's output.[5] The figure does not attempt to show finer detail than this, although there are a series of lesser climatic variations believed to be controlled by a series of cycles of progressively shorter periods. For example, a 200-year cycle may have resulted from smaller-scale solar variations.[6] In addition, there is the eleven year sunspot cycle, which amounts to a solar cycle with period an order of magnitude shorter.[7] Noncycling factors may also exist; a steady increase in atmospheric carbon dioxide has been reported in the 7,000-year interval from about 17,000 to 10,000 years ago; it may have been accompanied by a warming caused by the greenhouse effect.[8]

No attempt has been made to show the minor, short-term ups and downs of climate on the graph in fig. 1.6, as they are unlikely to have significantly affected the ecological developments that are the subject of this book. Many organisms, especially plants, lag in their response to climatic change.[9] It obviously takes considerable time for mature vegetation to become established on newly exposed ice-scoured rocks or glacial till (the unsorted material, ranging from clay to pebbles to boulders, deposited by an ice sheet). It also takes considerable time for whole ecosystems to change, with their numerous interdependent plant species, the habitats these create, and the animals that live in the habitats. Therefore, climatically caused fluctuations in ecological communities are a damped, smoothed-out version of the climatic fluctuations that cause them.

Several of the climatic intervals shown in fig. 1.6 have traditional names, which have been shown in the figure. A digression is desirable here, on the naming of geological time intervals. The largest type of interval that need concern us is known as a *period*. We live in the Quarternary Period, which is merely a name for the current glacial age; it began between one and a half and two million years ago. Examples of earlier periods are the Tertiary Period, which immediately preceded the Quarternary and began about sixty-five million years ago. Before that was the Cretaceous Period, which began about 140 million years ago, and so on back into the more distant past. Periods are divided into geologic *epochs*. For example, the Tertiary Period is made up of five epochs (Paleocene, Eocene, Oligocene, Miocene, and Pliocene). The Quarternary Period so far has two, the Pleistocene and the Holocene. It would take us too far from our topic to delve into the reasons for these divisions.

The past 20,000 years, the time interval we shall be considering, brackets the end of the Pleistocene Epoch and the beginning of the Holocene Epoch, the epoch in which we live. In the light of modern knowledge, the division of the Quarternary into two epochs now seems unjustified. The transition between them, the Pleistocene/Holocene boundary, marks the end of the last glaciation, known as the Wisconsin glaciation (sometimes called Wisconsinan), and the beginning of the present interglacial. There seems no reason to doubt, however, that a new glaciation will soon begin, in which case there will be no reason in the future to pick out our present interglacial from the twenty-odd others that have interrupted the Quarternary Period (our glacial age) and treat it as in any way noteworthy. The only thing that distinguishes the Holocene Epoch from earlier interglacials (which are not treated as separate epochs) is that it includes *our* lifetimes, a thoroughly anthropocentric reason. Some geologists consider that the Holocene should not be recognized; they prefer to consider the Pleistocene as still going on. Among those who do favor the recognition of both Pleistocene and Holocene (the majority), there is a certain amount of disagreement over where the boundary between them should be put,[10] but the "official" date is 10,000 years ago.[11]

As figure 1.6 shows, comparatively warm intervals alternated with comparatively cool ones throughout the last 20,000 years. A list of names exists[12] for those in the first 10,000 years (the final years of the Pleistocene), but they are not in common use among paleoecologists and so are not given here. However, the intervals into which the second 10,000 years (the Holocene so far) have been divided have commonly used names that are shown in the figure. Precise boundary marks have been deliberately avoided for two reasons: first, beginnings and endings would be arbitrary, since climatic change was gradual; second, as we shall see in later chapters, these intervals were not simultaneous from one region to another. The correspondence between climatic changes in northern North America and factors affecting the whole earth need not have been close. Global temperatures would naturally have influenced the ice sheets, but their responses would have been delayed, since huge volumes of ice take a very long time to melt. Local climates would have been strongly affected by the extent of the ice sheets, which would have governed the temperatures of overlying air masses and consequently wind strengths and directions.

The first of the important ecological intervals in the Holocene is the hypsithermal interval, the name given to the warmest time interval in our interglacial, now in the irretrievable past. It came immediately after the rapid warming that marked the transition from the Wisconsin glaciation to the subsequent interglacial and, after lasting for 3,000 to 4,000 years, gave way to the Neoglaciation,[13] which was presumably the first stage of the coming glaciation. The cooling trend has not been continuous, however; the climate moderated for a while, producing the Little Climatic Optimum, and the cold then returned, to give the Little Ice Age. These two most recent intervals are a part of recorded human history; the rigors of everyday life during the Little Ice Age form an interesting chapter in the social history of the past 800 years.[14]

The Little Ice Age came to an end with the climatic warming that began in about the middle of the nineteenth century; this warming persisted until the 1940s, at which time a cooling trend set in that seems to have reversed itself about 1970. But these recent events, fascinating though they are, must not detain us. The time interval covered in this book begins about 20,000 years ago, when the ice sheets of the most recent glaciation, the Wisconsin, reached their greatest extent in North America, and ends a few hundred years ago, when European colonists began to introduce exotic animals and plants to the hitherto isolated continent.

The Dating Method

Another digression is necessary here to explain how dates are given throughout the book. The dates of events are always given as so many years before the present, abbreviated to B.P. The only weakness of the system is that, in theory, all dates must be adjusted continually to allow for the forward march of the present; adjustments are unnecessary in practice, since the events that concern us are so far in the past.

There are a variety of ways of writing a date in the B.P. system. No one method has become standard, as there is no risk of confusion. For example, the date of an event that happened (or, rather, is thought to have happened) 12,600 years ago may be given as "12,600 yr B.P.," or "12,600 B.P.," or "12.6 ka B.P." (*ka* is an abbreviation for *kiloanni*, or 1,000 years), or "12.6k B.P." The last of these systems is the one used in this book because it is brief and clear.

The Ice Sheets

The map in fig. 1.2 shows, roughly, the pattern of the North American ice sheets as they were at the time of the late Wisconsin glacial maximum. They were probably around fifteen million square kilometers in area. Earlier maxima had occurred during the Wisconsin glaciation; climate is never constant and the margins of the ice sheets had advanced and retreated more than once during the preceding 60,000 to 70,000 years since the glaciation began. The late Wisconsin glacial maximum was the final maximum, however, after which the ice sheets as a whole began to shrink. Needless to say, the shrinkage has not been steady—there have been temporary, local readvances—but once the warming of the present interglacial began, the total quantity of ice dwindled rapidly.

The details of ice sheet geography are the subject of much debate.[15] For example, it is not known whether Nova Scotia and Newfoundland had their own small ice sheets or caps (as shown in fig. 1.2) or were merely overrun by lobes of the great Laurentide ice sheet. (The locations of places named in this and the subsequent three sections are shown on the map in fig. 1.7.) Nor is it known whether there was an unbroken Cordilleran ice sheet (as in the figure) or merely a complicated network of large mountain glaciers with numerous local ice fields like the modern Columbia Icefield; nor whether the Arctic islands were covered by a mosaic of separate ice caps (as in the figure) or an undivided ice sheet.

Ice sheets and ice caps (small ice sheets) are by no means inert. On the contrary, they grow when they are "fed" and waste away or become stagnant when they are "starved." As they grow, the ice flows. Movement is always going on somewhere in a well-nourished ice sheet, which is therefore deservedly called an "active" sheet. For an ice sheet to remain active, it must be constantly fed with fresh supplies of snow. Then, provided the summers are too cool for each winter's snow to melt away completely, the snowfalls of successive years steadily accumulate, the compressed snow at the bottom turns to ice, and the ice sheet (or cap) grows thicker. Its center builds up to form a dome that squeezes the underlying ice and makes it flow outward.

Therefore, an ice sheet spreads, and the spread is ultimately limited by one of two causes. Either the margin of the sheet reaches a region where summers are warm enough to melt the ice as fast as it arrives (the margin then becomes stationary, at least temporarily, and marks a line of dynamic equilibrium where the rate of melting just balances the rate of inflow), or the margin reaches the coast and flows beyond it to form an ice shelf over the sea. The shelf may or may not be grounded at its inner edge, depending on the depth of the water just off shore, but at least the outer part of the shelf floats, and its seaward edge is constantly eroded by the action of the sea.

Blocks of ice repeatedly break away in the form of icebergs; the ice shelf is said to produce bergs by "calving." At the same time, because of the comparative warmth of sea water, the ice shelf constantly melts at its outer cliff face. Once again, the margin of the ice represents a line of dynamic equilibrium, where the rate of wasting just matches the rate of advance of the shelf.

Now consider any small ice-covered region near the outer edge of the whole glaciated area. The problem of whether the ice covering the region was a detached, local ice cap or merely part of the main ice sheet is of much more than glaciological interest. It has important ecological implications. Suppose the local ice is known to have expanded at some past time. If it was a small, independent ice cap, it follows that climatic conditions in that locality must have changed in a way that favored growth of the cap. The local temperature must have declined, or the local snowfall increased (or both), and the ecosystems just beyond the edge of the ice must have been affected. But if the ice in the region were merely the outer part of a big ice sheet, its expansion could have been due to climatic change thousands of kilometers away at the center of the sheet, causing a buildup of the ice there. Once a growing dome of ice became so high that its slopes were steep enough to be unstable, it would have given way at the margins. Lobes of ice would have surged rapidly forward, pushed by the pressure behind, even in the absence of any climatic change in the region where the expansion occurred. Full-grown forests, growing in a climate hospitable to trees, could then have been overrun and crushed by the surging ice lobes.

Ice sheets are more than simply thick, flat layers of ice. They have interesting topographies. Although the surface of a big ice sheet would have looked flat to an observer standing on it, it would in fact have been undulating, with enormous, gently sloping hills (domes) separating by equally enormous and gently sloping passes (saddles). Moreover, as the climate warmed and the ice melted, the domes and saddles slowly migrated.[16] The largest ice sheet, the Laurentide sheet, probably had a single dome centered over Hudson Bay some time before glacial maximum, which was about 20k to 18k B.P.; ice thickness at the center may have been as much as five kilometers. The ice sheet, as it melted, is believed to have collapsed at the center, under the dome. Two separate domes resulted, a Keewatin dome and a James Bay dome, with a saddle between (see place names in fig. 1.7). As melting continued, the James Bay dome sank in its turn, widening the saddle and leaving a dome over Quebec and Labrador. Finally, between 8.5k B.P. and 8k B.P.[17] the saddle melted away entirely, leaving the domes it had separated as two independent ice sheets (see fig. 1.5); these melted away in their turn, with the Quebec-Labrador sheet lasting longer than the Keewatin sheet. By about 6.5k B.P.[18] the great Laurentide ice sheet had vanished except for segments in the arctic islands that still persist.

It is noteworthy that melting was not confined to the southern edge of the ice. Shrinkage occurred all around the periphery of the great Laurentide sheet and the two separate sheets that succeeded it, with the northern margins of the ice retreating southward at the same time as the southern margins retreated northward. Thus, as the two final ice caps shrank, their margins moved inland from the Arctic Ocean shore, leaving a widening ice-free coastal strip (fig. 1.5).

These changes did not happen at a steady pace. The disappearance of the ice was brought about by two quite distinct mechanisms— "starvation" and melting. Starvation happened whenever snowfall was insufficient to feed the ice sheets. Melting took place independently. It is possible that the rate of melting varied in concert with the 2,500-year cycle in the sun's radiation if we assume that the cycle has been in progress at least since the time of glacial maximum. The dates at which warming and cooling "should" have happened if they were in accord with this cycle appear to coincide fairly well with the dates when melting is believed (on geological evidence) to have speeded up and slowed down. There were rapid warm-ups of climate, accompanied by rapid melting of the ice at around 14k B.P.[19] and 10k B.P.[20] The earlier of these occurred while the Cordilleran ice sheet existed, and the warming affected it as well as the Laurentide sheet, which suggests that a worldwide climatic change was the cause. These warm-up periods alternated with contrasting periods in which climatic warming, and the consequent melting of the ice, slowed or even reversed. For example, from about 13k to 10.5k B.P. there was a distinct pause in the warming trend; the southern margin of the Laurentide ice shifted back and forth repeatedly within a zone only a few hundred kilometers wide.[21] Again, not long before the onset of the warm hypsithermal interval that lasted from about 8.5k B.P. to about 5k B.P., there was a so-called "still-stand," a pause in the melting of the ice sheets that lasted for a few centuries.[22]

It is interesting that dwindling ice sheets were present throughout most of the hypsithermal, which was (presumably) the peak of our current interglacial. Huge masses of ice take a long time to melt, and they lag far behind climatic change. The final disappearance of the mainland ice sheets probably did not happen until 6.5k B.P.; ice caps in the Arctic islands still have not disappeared. Very probably they will not; there will not be sufficient time before the ice sheets of the next glaciation begin to grow.

Ice and Sea

While all this was happening, equally spectacular changes were going on at sea and around the coast. At the height of the last glacial maximum so much of the world's water was locked up in continental ice sheets that the worldwide sea level was at least 85 meters below its present level, and perhaps as much as 130 meters below. Opinions differ.[23] Huge tracts of what are now parts of the submerged continental shelves of North America were then dry land.

From the paleoecological point of view, by far the most important of these tracts was the wide land bridge that linked present-day Alaska with present-day northeastern Siberia (see fig. 1.2). It provided a corridor between Eurasia and North America for a large fraction of the sixty or seventy millennia of the Wisconsin glaciation and was finally broken about 15.5k B.P., when rising waters joined the Pacific and Arctic oceans, thus creating the modern Bering Strait.[24] The land bridge and lands immediately to east and west of it have acquired the name *Beringia*, which is an extremely important geographical entity in the ecological history of North America.

The melting ice sheets and the rising sea interacted with each other to produce some interesting positive feedbacks.[25] Where the margins of the ice sheets reached the ocean and formed floating ice shelves, the rising sea hastened the calving of icebergs. Besides being diminished by the calving off of bergs, the ice sheets were affected in yet another way. The melting bergs cooled the sea and led to reduced evaporation of water vapor from the surface. This in turn led to reduced snowfall, which eventually starved adjacent ice sheets. At the same time, a large quantity of fresh meltwater must have been flowing off the ice sheets into the sea. It would have spread out to form a thin layer of fresh water on top of the saltier, and therefore denser, sea water and, being fresh, would have frozen readily in winter. The resultant thin layer of ice would have formed a lid over the sea, reducing the amount of water vapor available to form snow. The effects of melting icebergs in summer, and an ice layer over the sea in winter, reinforced each other in hastening starvation of the ice sheets.

Note that the level of the sea is not necessarily closely related to the area of the ice; it is the volume of the ice, not its area, that counts. It has been estimated that the volume of the ice sheets decreased rapidly between 16k and 13k B.P., and sea level rose accordingly.[26] The area of the ice did not decrease markedly in this interval (compare figs. 1.2 and 1.3), but it did get much thinner.

Sea level can rise and fall in two quite different ways: there are *eustatic* changes and *isostatic* changes in sea level. A eustatic change affects the whole world ocean and results from a change in the world-wide volume of sea water. An isostatic change is a local change resulting from local warping of the earth's crust. To put it another way, a eustatic change is a rise or fall of the sea relative to the land, and an isostatic change is a rise or fall of the land relative to the sea. The preceding paragraphs described eustatic changes; however, as the ice melted, a variety of isostatic changes occurred as well.

The isostatic changes were chiefly caused by the depression of the earth's crust under the weight of the overlying ice sheets (but see The South Coast of Beringia, Chapter 6); as the ice melted, the exposed tracts of land were lower than contemporary mean sea level. In two places sea water flooded in, bringing the sea to what are now inland areas. The first of the big inland seas to form was the Champlain Sea (see fig. 1.4), which was in the valleys now occupied by the Saint Lawrence and Ottawa rivers. Later, after the Laurentide ice sheet had split into two parts, the sea flooded into the Hudson Bay basin; as the ice margins to east and west receded, the new sea expanded. At its largest it appeared as in figure 1.5. It is known as the Tyrrell Sea and was a forerunner of modern Hudson Bay, which it greatly exceeded in size.

Both the disappearance of the Champlain Sea and the shrinking of the Tyrrell Sea were caused by the rebound of the earth's crust once it was relieved of the load of ice that had been pressing it down. Land continues to rise long after the weight depressing it has been removed. Because of this, rebound is still going on, especially in the Hudson Bay region, where the most rapid uplift in the world is now occurring on the mainland shore opposite the Belcher Islands. Isostatic uplift there is proceeding at a rate of fifteen millimeters per year,[27] and in time Hudson Bay will drain away completely[28] unless the growth of new ice sheets halts the process. Repeated sea level changes throughout the past 20,000 years have thus affected the whole coast line of glaciated North America. Ever since melting of the ice began, at the end of the final Wisconsin glacial maximum, eustatic and isostatic changes have been going on simultaneously so that what happens at any given locality is the result of two processes. And there is an added complication: isostatic adjustments do not invariably cause sea level to fall. Just beyond the margin of an ice sheet that is depressing the crust beneath it, a compensating upward bulge forms. Then, when the ice sheet shrinks, the bulge (an isostatic upwarping) migrates in the wake of the receding ice margin and a crustal subsidence (an isostatic downwarping) follows in the wake of the bulge. The outcome of this process at any one point is first a rise in sea level (relative to the eustatically determined level) when the ice

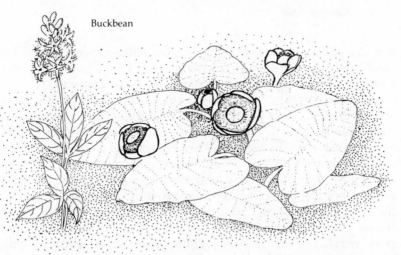

Buckbean

Spatterdock

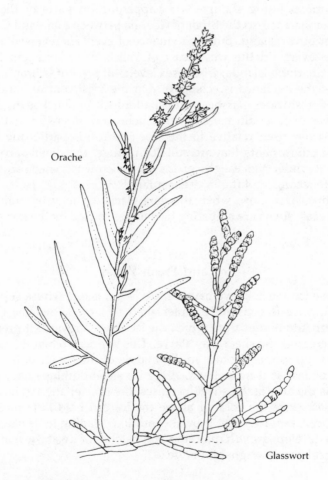

Orache

Glasswort

first disappears, followed by a fall as the bulge arrives, followed by another rise once the bulge has gone past. These processes are happening now on both sides of the North American continent.

In the east, the Maritime provinces of Canada (Nova Scotia, New Brunswick, and Prince Edward Island) are sinking (or equivalently, the sea around their shores is rising[29]) as a depression in the surface follows in the wake of a bulge migrating to the north. Some of the evidence is ecological. At this very moment, the rising sea is inundating spruce forest and killing the trees; the forest is succeeded first by brackish-water marsh dominated by the seacoast bulrush, then by tidal salt marsh dominated by cord grass and salt hay. Finally, the sea takes over, but under the sand can be found buried cord grass peat, which overlies seacoast bulrush peat, which in turn overlies freshwater peat with buried, upright spruce stumps.

Corresponding changes are happening on parts of the west coast, for instance, in the Strait of Georgia between mainland Canada and Vancouver Island. But the sequence of events is reversed a short distance away, on the west coast of Vancouver Island and in the Queen Charlotte Islands, where sea level is at present falling.[30] Again some of the evidence is ecological. On the west coat of Vancouver Island,[31] freshwater marshes, with spatterdock and buckbean, overlie what was once a salt marsh, with orache and glasswort; evidently, the land has risen relative to the sea. The rise is partly due to the isostatic adjustments that are still "correcting" the downwarping of the crust under the weight of the Cordilleran ice sheet and also, partly, to tectonic uplift of a crustal plate.[32]

Thus, even now, when the next glaciation may be imminent, coastal ecosystems are adjusting to changes caused by the last glaciation.

Ice and Fresh Water

An active ice sheet, as we have already seen, is one whose supply of snow and ice is constantly replenished by new snowfalls; the ice builds up into domes from which the ice flows outward. If precipitation decreases, the sheet may starve. This happens when the ice that has built up to form domes in the middle of the ice sheet is not replaced as fast as it flows away. A big ice sheet develops its own climate; as the area of the sheet increases, the precipitation at its center decreases, since the cold air above the ice can hold little moisture. Therefore, as an ice sheet grows, conditions leading to its starvation tend to develop; growth may stop and patches of ice at the margin of the sheet become stagnant, or "dead."

Large tracts of stagnant ice existed along the southern and southwestern margins of the Laurentide ice sheet for long periods during the final years of the Wisconsin glaciation. When the ice front was retreating across the region now covered by the prairie grasslands of the Midwest, some big islands of stagnant ice were left behind.

The Laurentide sheet did not shrink without interruption, however. There were a number of readvances during which lobes of ice flowed south. For example, between 14k and 10k B.P., a succession of at least four advances flowed into Iowa, South Dakota, the Lake Michigan basin, and the Lake Superior basin.[33] Depending on the climate at the time, advances of the ice were sometimes over frozen ground and sometimes over unfrozen ground. The state of the ground determines the way in which ice flows. When it flows over frozen ground, ice oozes forward by a process of continuous bulging at the front; that is, it "creeps," undergoing continuous internal deformation as it does so. When it flows over thawed ground, however, besides creeping, it also slides; the ground warms the base of the ice and forms a film of meltwater that acts as a lubricant.

Sometimes, readvancing ice flowed back over masses of stagnant ice that had been left behind. When the upper layer of ice subsequently wasted away, it deposited a huge quantity of glacial drift that insulated the stagnant ice beneath and delayed the melting of this lower layer for millennia. In the Missouri Coteau region of North Dakota, for example, it took about 3,000 years, from 12k to 9k B.P., for a buried sheet of stagnant ice to melt. At the outset the sheet was probably 100 meters thick.[34]

The insulating layer of drift was made up of rock fragments, ranging in size from huge boulders to the finest "rock flour," that had been carried on, in, or under the readvancing ice. Some parts of the layer consisted of *till*, that is, drift let down in place when the ice melted. Other parts consisted of *outwash*, that is, drift carried in torrents of meltwater and deposited beyond the ice front.

Numerous ponds and lakes formed in stagnant ice country, and their ecology is described in chapter 8. Reminders of the time when stagnant ice was common are still with us in the form of "prairie potholes," the numerous freshwater ponds of the midwestern plains. Most of them occupy *kettleholes*, depressions left in the ground when isolated, long-lasting blocks of stagnant ice finally melted. Prairie potholes contribute enormously to the ecological diversity of the plains; they also provide resting places for vast numbers of migrating ducks and nesting sites for ducks that breed in the region. They account for fifty percent of duck production in North America in an average year.[35]

To return to the topic of the Laurentide ice sheet and the recession of its southern margin. Warming proceeded, notwithstanding temporary interruptions. The result, of course, was enormous volumes of meltwater. The plains of the Midwest slope upward, very gently, from east to west. The ice front was receding downslope toward the east-northeast; therefore, the meltwater was ponded between ice on one side and rising ground on the other. Enormous *proglacial* freshwater lakes formed, comparable in size to the modern Great Lakes. When storms blew up over these freshwater inland seas, the waves must have eroded their ice shores. Wherever an ice cliff was exposed to rising water levels and wave action, bergs were no doubt calved off. Thus in spite of the great differences in terrain between the lands on the northern and southern sides of the ice sheet, the scenery to north and south may have been rather similar in places; in summer, there would have been expanses of stormy water dotted with recently calved icebergs and with an ice cliff forming one shore.

The geography of the region south and southwest of the Laurentide ice changed constantly, exhibiting a continuously shifting pattern of ice, land, and water.[36] Because the land surface is so nearly

Ice-front Lake

Glacial Spillway

level, it takes only a slight rise or fall in water level to produce a pronounced change in a lake's outline. Figures 1.3 and 1.4 show speculative reconstructions of the arrangement of the large proglacial lakes on two occasions separated by 3,000 years. The change from the earlier pattern to the later should not be thought of as having happened gradually and smoothly; on the contrary, conditions close to the melting ice front must have changed in unexpected ways almost from day to day.

In addition to the large lakes shown on the maps, there were many thousands of smaller, shallower, even shorter-lived lakes, also dammed by the ice. As the ice melted, its receding front uncovered a succession of hitherto buried hills and valleys, ridges and slopes, saddles and hollows. Although the plains undulated so gently that these topographical features were of low relief and inconspicuous, they nevertheless controlled the direction of flow of the meltwater. The uncovering of a valley must have often converted what had been a lake into a swiftly flowing river. Indeed, the ice front itself was a topographical feature, albeit an impermanent one. As it shifted, it changed the shapes of drainage basins and the direction in which water flowed from them. Deeply eroded overflow channels were suddenly left high and dry when a new drainage route, in a new direction, was uncovered. Many of these abandoned spillways, known as coulees, occur in the western plains.[37]

The formation and disappearance of numerous temporary lakes strongly influence the northward spread of land plants as the ice disappeared. The large lakes, especially, were barriers to migration, but whenever a lake drained, its exposed bed provided a tract of new land for colonization by pioneer plants. The frequent disruption and rearrangement of drainage systems had interesting biogeographical consequences. For example, many streams and rivers draining the eastern slopes of the Rocky Mountains in what is now Alberta flowed into the Milk River and thence into the Missouri when the lowlands to the east were ice covered. Now the same rivers drain into the Saskatchewan River system, which flows into Hudson Bay. This is only one of several such shifts that greatly affected the present-day distribution of aquatic organisms, as we shall see in later chapters.

Ice and Atmosphere

Preceding sections describe, in general outline, the ice-land-freshwater-seawater system of glaciated northern North America. There is one other exceedingly important component of the total system to consider, the atmosphere. Two characteristics of the atmosphere are of special importance to ecologists and paleoecologists: its movement, which is wind, and its ability to hold moisture, which determines precipitation.

First, the winds. When the ice sheets were big, they must have been the source of strong *katabatic* winds. A katabatic wind is the flow of cold and therefore dense air down a sloping surface of land or ice; strong katabatic winds flow off the world's two existing ice sheets, in Greenland and Antarctica; at one antarctic station a mean hourly wind speed of 156 kilometers per hour has been reported.[38] It seems certain that similar winds blew off the North American ice sheets. The summits of the ice domes were probably at an altitude of about 3,500 meters above sea level.[39] Contact with the ice cooled the air above them, and smooth slopes fell away in all directions. The conditions would therefore have been ideal for fierce katabatic winds.

Such winds are cold, but not excessively so. When the air begins its descent from a high summit it may indeed be exceedingly cold. But as it descends, its pressure increases and its temperature rises automatically, by *adiabatic* warming. (An adiabatic temperature change is one due entirely to altered air pressure without any exchange of heat between air and environment. Chinook winds are also warmed in this way; a chinook is much warmer than a katabatic wind, however, since it starts out warm.) By the time the katabatic winds had reached the margin of the ice sheet, they had tremendous force but probably were no colder than winter winds in the same region at the present day.[40]

These gales blew over recently deglaciated land, where vegetation was absent or at best very sparse; the ground surface consisted of newly exposed glacial till containing quantities of ground-up rock ranging from coarse sand to fine rock flour. The wind picked this material up, and the resultant sand storms and dust storms must have far surpassed the worst that occur at the present day. The wind deposited this material in the form of sand dunes and in extensive, thick beds of dust known as *loess*. These were the surfaces available for colonization by plants when seeds were blown in from the south, or from east or west along the ice front. There may well have been a windswept corridor between forested land to the south and ice sheets to the north along which seeds and other small organisms could be airborne for great distances. Wind was probably even more important as an ecological agent then than now in the central plains of the continent.

We must next consider the effects of the ice sheets, and of their melting, on precipitation patterns. The chief cause of change in precipitation patterns was the constantly varying geographical pattern of land and sea. At glacial maximum, sea level was low, land that is now submerged was exposed (see Ice and Sea and fig. 1.2), and there were probably no big inland lakes. Then, as the ice melted, sea level rose, and a succession of big proglacial meltwater lakes came into existence (see Ice and Freshwater and figs. 1.3, 1.4, and 1.5). The result of these changes was a continuously shifting pattern of local climates.

Near large bodies of water (salt or fresh) a maritime type of climate prevailed; there was enough moisture for abundant rain and snow, and the annual range of temperature variation was damped by the tremendous capacity of the sea or large lakes to absorb and retain the sun's heat. Away from the big water bodies, the climate was continental, with low precipitation and temperatures ranging from extremely high in summer to extremely low in winter. As the climate at any place changed in response to the shifting of ice sheets and water bodies, corresponding ecological changes must have occurred. For example, the interior of Beringia was far from any marine influence when sea level was at its lowest, and it presumably had an arid continental climate. When the sea rose and Bering Strait came into existence (see fig. 1.3), the climate of the peninsula extending west from present-day Alaska changed from continental to maritime, with a corresponding vegetation change (see The Sundering of Beringia, chapter 10).

The topography of the big ice sheets also affected climate owing to the existence of *ice divides*. An ice divide is a height of ice from which ice flows away in opposite directions, in the same way that a *land divide* (such as the continental divide of North America) is a height of land from which rivers drain away in opposite directions.[41] In the case of an ice divide, of course, both the divide itself and the "rivers" flowing down its slopes are made of the same material, ice.

An ice divide is a connected row of domes and saddles in an ice sheet in the same way that a land divide is a connected row of mountains and cols. Both kinds of divide strongly influence climate. The climatic effect of land divides such as the several north-south trending mountain ranges of western North America is well known. These divides force moisture-laden winds off the Pacific to gain altitude and precipitate their moisture as rain or snow. As a result, winds that were initially moist are converted to dry winds, and the effect is made visible by the contrasting types of vegetation on the two sides of the divide. The dry side is often said to be in the "rain shadow" of the divide.

In exactly the same way, the ice divides of the big ice sheets,

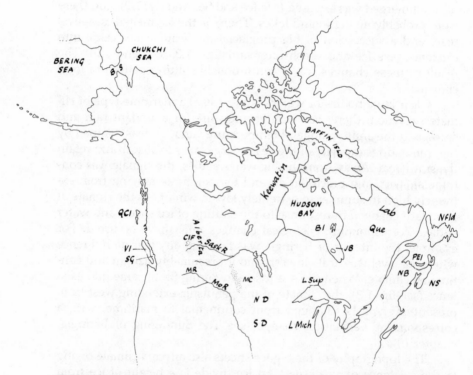

FIGURE 1.7: Map showing places referred to. BS, Bering Strait; QCI, Queen Charlotte Islands; VI, Vancouver Island; SG, Strait of Georgia; CIF, Columbia ice field; Sask R, Saskatchewan River; MR, Milk River; MC, Missouri Coteau; Mo R, Missouri River; ND, North Dakota; SD, South Dakota; L Sup, Lake Superior; L Mich, Lake Michigan; BI, Belcher Islands; JB, James Bay; Lab, Labrador; Que, Quebec; NB, New Brunswick; NS, Nova Scotia; PEI, Prince Edward Island; Nfld. Newfoundland.

which trended roughly west to east, must have created "snow shadows" on their northern slopes. The prevailing winds were from the south. Interesting feedbacks may have developed. As the snow accumulated and turned to ice, the large quantities of snow falling on the side of an ice divide exposed to moist winds would have caused the summit of the divide to grow steadily higher while simultaneously it migrated to windward (i.e., southward). Because of the increasing height of the divide, the snowfall rate in its shadow would have grown steadily less, the ice in the shadow would have become starved, and the ice sheet as a whole would have shrunk. This process may have caused the northern margin of the Laurentide ice sheet to retreat at times when its southern edge was advancing.

It is not known whether advances and retreats of the ice on the northern and southern sides of the sheet alternated with each other or were synchronous. If they were caused by the feedback process just described, they would have alternated. If, as is quite possible, they were caused by worldwide temperature variations, they would have been synchronous. Interestingly, although the feedback mechanism entails pronounced regional changes in precipitation rates, it need not have caused temperature changes. Conversely, if temperature variations caused the advances and retreats, precipitation rates did not necessarily change.

It should now be clear that the physical environment of northern North America has changed dramatically in the past 20,000 years. This environment was (and still is) a complicated and intricate system, powered by the sun and having land, ice, freshwater, salt water, and atmosphere as its components. Even if it were lifeless, it would still be dynamic; the components would continue to interact with one another. One of the most interesting aspects of this never ending change from the ecological point of view is that, over the time interval we are considering (and probably for the whole of the earth's history), physical conditions on this continent (and everywhere else) have never repeated themselves.

At no time has there been a return to "things as they were." It is true that there must have been times when average temperatures were similar to those of the present. Thus, before the beginning and after the end of the warmer-than-now hypsithermal interval, the average annual temperature must, for a while, have been much the same as now. But in other respects, conditions would have been radically different, as there were still extensive ice sheets that would have cooled their immediate neighborhoods, and sea level was still about twenty to twenty-five meters lower than at present.[42] Nor is it reasonable to assume that conditions on the ice sheets were the same

Nunataks

as those in Greenland and Antarctica today. The North American ice sheets were at a much lower latitude; they did not experience months of winter darkness, and the altitude of the midday sun in summer was much greater.

Ice-free Land: Refugia and Nunataks

Every living thing to be found in the once ice-covered areas of North America must, obviously, be descended from ancestors that, at glacial maximum, were living in an ice-free area. (Or nearly every living thing. Exceptions to this sweeping statement are the few cave invertebrates believed to have survived in caves under the ice. See Refugia near the Ice-Free Corridor; chapter 7.) The whereabouts of ice-free areas (or *refugia*), what the conditions in them were like, and which species survived in them are topics of endless fascination to paleoecologists.

As a glance at fig. 1.2 shows, three large regions served as the chief refugia for life at the time of maximum Wisconsin glaciation. They were, first, midlatitude North America, that is, all the land south of the ice and including the so-called ice-free corridor that formed a northward-pointing peninsula of ice-free land just east of the Rocky Mountains; second, Beringia, the enormous tract made up of what are now most of Yukon and Alaska together with the beds of two shallow seas—the Bering Sea and the Chuckchi Sea—and the easternmost peninsula of Siberia (see fig. 1.7); and, third, the coastal plains east of the ice sheet, which are now submerged to form the continental shelf off the shores of New England and the Atlantic provinces of Canada.

In addition to these big refugia, there were almost certainly many small ones, some only a few square kilometers in area and too small to be shown on small-scale maps of the ice sheets. The small refugia were of two kinds: *nunataks* and *coastal refugia*.

A nunatak is a mountain summit that protrudes through an ice sheet. The word is used in two senses. It can mean a summit that was surrounded by ice during part or all of the Wisconsin glaciation but was never overridden. This is the definition used in this book, unless specifically stated otherwise. The word is also used to mean a summit protruding through the ice at the time under discussion, either now or in the past; a nunatak in this sense could have been ice covered at some other time.

Nunataks were probably common in the mountainous regions of glaciated North America. When an ice sheet formed and grew, simultaneously thickening and creeping forward, it must have risen up mountain slopes in its path in the same way that a swelling flood of water would rise, though at "glacial" speed of course. Many summits would be submerged, but the highest ones would remain exposed, above the general level of the ice. Nunataks have therefore existed as open ground for at least 100,000 years, that is, since the beginning of the interglacial (known as the Sangamon interglacial) that preceded the Wisconsin glaciation; they may have been free of ice during earlier glaciations as well.

Judging whether a mountain peak was once a nunatak is often difficult. Several kinds of evidence have to be considered.[43] For example, a deep layer of *felsenmeer* on a summit is taken by many geomorphologists as evidence that the summit was a nunatak. Felsenmeer is the chaotic mass of broken rocks that steadily accumulates on high summits as the bedrock is fractured by alternate freezing and thawing. It takes time to form, and there is a limit to the amount that could have accumulated on a summit that had been scoured by Wisconsin ice. The evidence is thought to be particularly strong if felsenmeer is deep above a certain contour on a mountainside and noticeably thinner below it; this suggests that the contour marks the topmost level of the ice that once engulfed the mountain. Negative evidence also carries weight; absence of glacial striae (scratches made by moving ice) on the rocks suggest that a summit was a nunatak.

But the presence on a summit of a thick layer of felsenmeer, and the absence of signs of ice scouring, can also be interpreted to mean that the summit *was* covered by ice; not by moving ice but by stationary ice, frozen fast to the underlying rock. Such landfast ice would have left no scratches and would have preserved, undisturbed, any "old" felsenmeer (dating from before the last glaciation) lying on its surface. The old felsenmeer on lower slopes, on the other hand, would have been partly swept away by warmer, moving ice.

Although it is often debatable whether a specific summit was a nunatak, the evidence strongly suggests that they were fairly common. In spite of the difficulties of recognizing them, many nunataks are known or suspected. The maps in fig. 1.8 show those that will be mentioned in later chapters.

Two sets of circumstances would have made nunataks quite numerous. First, they would have occurred wherever mountains, or even quite modest hills, lay within the margin of an ice sheet that was prevented from thickening and advancing by a lack of snow and

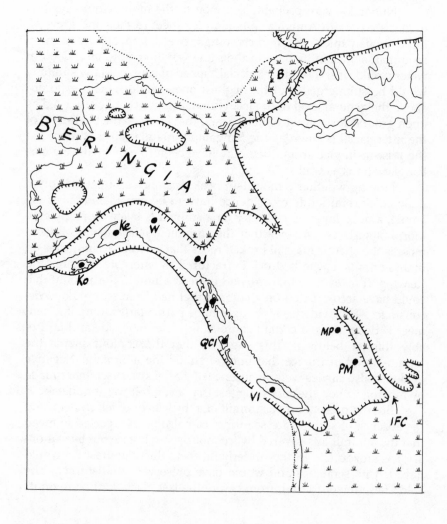

hence starvation. An ill-nourished ice sheet would have been thin at
its borders. Second, they would also have occurred where high
mountains lie close to deep ocean with only a narrow strip (if any) of
coastal plain and continental shelf between[44] (fig. 1.9). An ice shelf
cannot project over the deep sea beyond the continental shelf because winds and currents break it up; there is a limit to the possible
height of the ice cliffs at the margin of a sheet and there is a limit, if
an ice sheet is stable, to its possible steepness of slope. The upward
gradient, as one goes from the edge of the ice toward the continental
interior, is usually about 1:200 and almost never exceeds 1:100. Therefore, mountains close to the seaward margin of a stable ice sheet can-

FIGURE 1.8: The sites of thirteen important nunataks (black dots) and two
coastal refugia. The nunataks are (in the western map) Ko, Kodiak Island;
Ke, Kenai Peninsula; W, Wrangell Mountains; J, Juneau ice field; A, Alexan-
dra Archipelago; QCI, Queen Charlotte Islands; VI, Vancouver Island; MP,
Mountain Park; PM, Plateau Mountain. (In the eastern map): TM, Torngat
Mountains; SS, Shickshock Mountains; CB, Cape Breton Island; LR, Long
Range Mountains. The coastal refugia (both in the eastern map) are G,
Gaspé Peninsula; N, Newfoundland coast. Other symbols: B, Banks Island;
IFC, ice-free corridor; G St L, Gulf of Saint Lawrence.

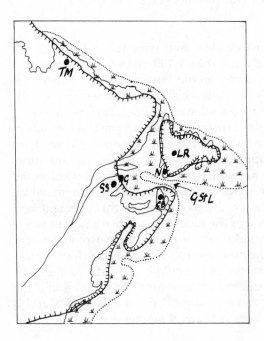

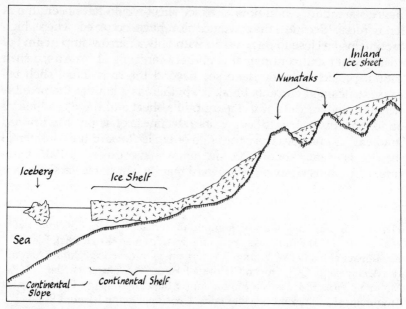

FIGURE 1.9: Nunataks in coastal mountains. The vertical scale in the diagram is greatly exaggerated. Because there is an upper limit to the slope of an ice sheet's upper surface, high summits close to the edge of the sheet must protrude through the ice. (Adapted from Dahl 1946.[44])

not be ice covered. If the marginal ice cliffs were 100 meters high, for example, most summits of more than 1,100 meters elevation within 200 kilometers of the ice margin would be nunataks. But note that this argument applies only to stable ice sheets; surface slopes can be much steeper than 1:100 near the edge of actively growing sheets.

As already mentioned, nunataks were not the only small refugia. There were also small coastal refugia. These were at sea level and were, in a sense, complementary to the coastal-mountain nunataks just described. They were short sections of coastal plain in the lee of high mountains. The glaciers flowing down from an inland ice sheet through valleys between the mountains might, or might not, coalesce on the seaward side of the mountains. If two adjacent valley glaciers did coalesce, the mountain between them would be converted into a nunatak; if they did not coalesce, the stretch of ice-free shore between the two glacier snouts would be a coastal refugium.

Coastal refugia are believed to have been common around the coasts of Newfoundland[45] and the Gaspé Peninsula (labeled N and G respectively in fig. 1.8). There is much disagreement among glacial geologists about the locations of shore lines in the Gulf of Saint Lawrence region at the time of glacial maximum, and the "coastal refugia" of some specialists would be a long way inland on the maps of other specialists. In theory, it would take a series of maps, showing ice margins and shorelines at a sequence of times, to fix the positions of coastal refugia exactly, since they must have migrated slowly inland as the sea level rose and the ice front receded. The ecosystems occupying them would have migrated similarly; plant and animal populations would have spread inland while the seaward margins of the refugia were being inundated by the rising sea.

While geomorphologists disagree on which mountains were nunataks and which stretches of shore were coastal refugia, paleoecologists disagree on what the living conditions were like in refugia of all kinds. Subsequent chapters have much to say on this subject; for the moment we will consider only nunataks.

Modern nunataks, in Greenland and Antarctica, are bleak in the extreme. This makes it interesting to speculate about whether all ice

Coastal Refugium

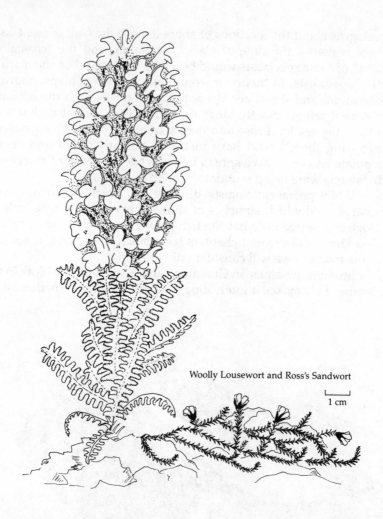

Woolly Lousewort and Ross's Sandwort

1 cm

age nunataks were similarly bleak or whether those at middle latitudes, which have no modern counterparts, were comparatively
balmy, at least in summer when the midday sun was as high in the
sky as it is nowadays. Nobody doubts that conditions must have been
harsh on nunataks at high latitude and far from the open sea. Note
the word *open*. Until the climate had warmed considerably, most of
the Arctic Ocean and even part of the North Atlantic were covered
by sea ice all through the year. Sea ice affects climate as land does; for
a maritime climate (moist and mild) to prevail, the nearby sea must
be ice free.

The more northern nunataks were therefore extremely arid, as
well as extremely cold; they would have been islands of arctic desert,

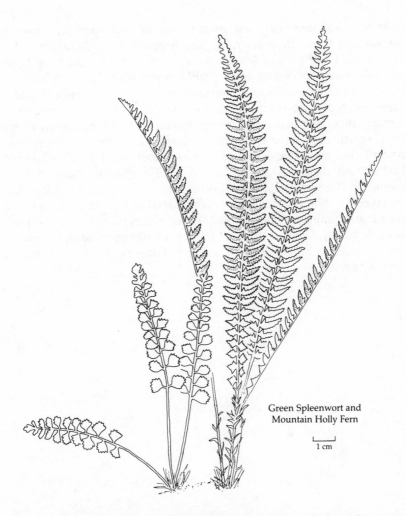

Green Spleenwort and
Mountain Holly Fern

|___|
1 cm

that is, stony tundra with much bare ground, where few plant species could survive. The Norwegian plant geographer Eilif Dahl[46] has drawn up a list of such plants, which he describes as "typical tundra refugees." Two examples are woolly lousewort and Ross's sandwort. Both are species found in dry, stony places; their present range includes parts of northernmost Greenland and the northeastern tip of Siberia, as well as the high arctic Canadian islands. They undoubtedly grew in Beringia for the whole of the Wisconsin glaciation and probably on some of the harsh climate nunataks as well.

Conditions on the coastal-mountain nunataks at midlatitude, for example, those on Vancouver Island and the Queen Charlotte Islands, were much less harsh. The open ocean was not far away, precipitation was plentiful, and sunny summer days were as warm as (or warmer than) they are now. The hospitable conditions (what a skier would call perfect spring skiing conditions) would have been ideal for many species of plant. Dahl has given a long list of what he describes as "typical coastal mountain refugees." They are mountain plants with a more southerly distribution than the tundra refugees. Among them are two ferns, green spleenwort and mountain holly-fern, which now have extensive ranges in the western mountains, being found most frequently at or near timberline. They also grow in eastern Canada and southern Greenland. Present-day plants of these species in the western mountains may be descendants of plants that survived the Wisconsin glaciation on coastal-mountain nunataks, but they may equally well be descendants of plants that survived in mid-latitude North America south of the ice. However persuasive the evidence seems to be that a plant is a nunatak survivor, the conclusion is never certain.

2

The Fossil Evidence

Fossils

To trace the ecology of the past, we must have evidence, and the most reliable evidence is provided by fossils. At first thought, one might think that fossils constituted the only evidence. How else can one infer what was living and where it was living thousands of years ago? Although it is true that fossils give the best, most incontrovertible evidence, we can draw interesting, if more speculative, conclusions about some organisms by mapping their present geographical ranges, an approach that will be discussed further in chapter 3. First, we will consider fossils.

Fossils are all about us, in uncounted millions. Although big fossils, such as the bones and teeth of large mammals, are fairly rare,

Fossil beetle fragments

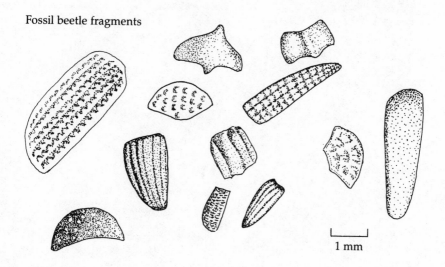

1 mm

fossils of microscopic size are abundant everywhere. They are unseen of course, except by the micropaleontologists who deliberately
search for them. It has become customary to divide fossils into two
classes, *macrofossils* and *microfossils*, depending on whether a microscope is needed to examine them.

Macrofossils exist in great variety. Usually only the hard parts
of organisms become fossilized; organisms consisting entirely of soft
tissues generally decay rapidly after death, leaving no trace of their
existence. Therefore, animals are most often represented by the

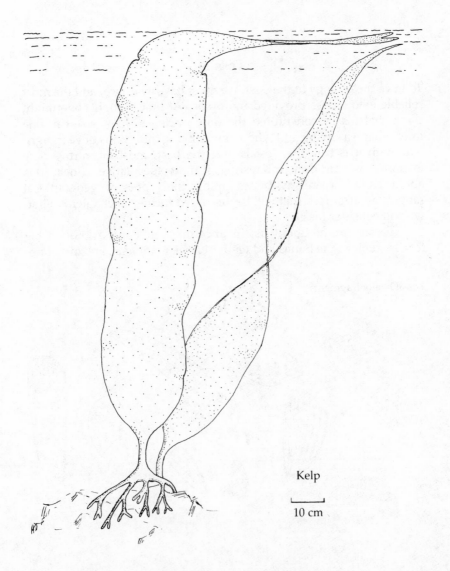

Kelp

10 cm

bones and teeth of vertebrates, the shells of snails and clams, and the hard parts of insects; these parts are usually broken into fragments. The commonest insect fossils are bits of beetles, especially pieces of the elytra (an *elytron* is one of the hard, upper wings of a beetle) and the pronota (the *pronotum* is the conspicuous "shield" covering most of the thorax). Identifiable beetle fragments can nearly always be found in organic sediments; they are especially well preserved in frozen soil.

Fossil leaves, twigs, wood fragments, seeds, and achenes (dry, one-seeded fruits) represent the plant community. Preserved pieces of burned wood, called fossil charcoal, give evidence of past forest fires. Soft plants or plant parts usually decay, leaving no fossil remains; an unusual example of an exception to this general rule is fossil fronds of kelp and other seaweeds buried in the sands of what was once a beach, but is now far inland. The beach was the shore of the Champlain Sea and is now high and dry in the Ottawa River valley.[1]

Microfossils

Microfossils are extraordinarily abundant and are enormously useful to research workers trying to reconstruct the past. Their abundance is not surprising since countless living microorganisms are everywhere in our environment, and many kinds have hard parts that do not decay when the organisms die. Thus microfossils remain in vast numbers as microscopic "skeletons," unseen except by the micropaleontologists who study them.

The most numerous of all terrestrial microfossils are pollen grains. Our knowledge of the history of vegetation is based almost entirely on the study of fossil pollen, together with fossil spores of nonflowering plants such as ferns and club mosses. The study of these microfossils is known as pollen analysis, or *palynology,* an important branch of micropaleontology.

The results of palynological investigations form a large part of this book, and the techniques of pollen analysis are therefore considered in some detail in the next two sections. This section gives an overview of the other common kinds of microfossils. Fossils of six groups of organisms—diatoms, foraminifera, dinoflagellates, ostracods, chironomids, and stoneworts—are especially important in providing information about the ecology of the past. We consider each group in turn.

Diatoms are microscopic single-celled algae that are found in enormous numbers in both fresh water and the sea. A few species require dry habitats; they live on rocks, leaves, or soil. Of the much more abundant aquatic species, some live in the plankton, the communities of floating, drifting organisms that inhabit nearly all bodies of water, fresh or salt. Other species belong to the benthos, the com-

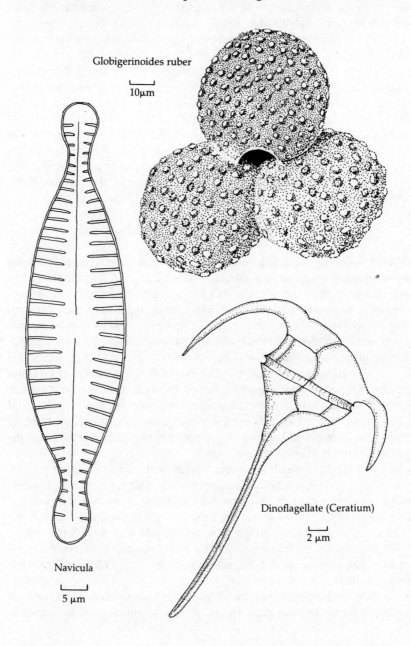

Globigerinoides ruber

10μm

Navicula

5 μm

Dinoflagellate (Ceratium)

2 μm

munity of organisms living on the bottom, again in fresh water or salt. In all diatoms, the living part of the cell is contained in a cell wall made up of two overlapping halves (valves) that fit together like a pillbox and its lid, or like the upper and lower parts of a Petri dish. The valves consist largely of silica, and it is this that makes them hard and highly resistant to decay.

There are thousands of species. Some are radially symmetrical, exactly like a Petri dish; others are bilaterally symmetrical and are often shaped like the specimen shown in the drawing, which belongs to the common freshwater genus *Navicula*. In all of them, the valves have intricately ornamental patterns, which provide the basis for their classification. Fossil diatom valves have accumulated in marine and lake sediments and serve as valuable evidence in paleoecological reconstructions. For many species, the habitats they required when they were alive are known, making it possible to deduce, from their presence as fossils, what the conditions in ancient habitats must have been like. For example, it is often possible to infer the acidity of an ancient lake from the species of diatoms found in the sediment.

Foraminifera (usually called *forams* for short) are another huge group of single-celled organisms. They are simple protozoa closely related to amoebas. Unlike amoebas, however, most of them have shells, and the shells are composed chiefly of lime (calcium carbonate). Each species has a shell of distinctive form, with characteristic surface ornamentations. These shells exhibit an extraordinary diversity of patterns; there are shells in the form of perforated spheres covered with sharp spines, shells like rows of beads fused together, shells like braided bread dough, and shells like miniature ammonites. The variety seems endless. The majority of species are marine, but a few live in brackish water and fewer still, in fresh water. Like diatoms, some live in the plankton and others in the benthos. The drawing shows a species belonging to the genus *Globigerinoides* from the ocean plankton. As with diatoms, so with forams: the species found in a sediment reveal much about the ecological conditions that prevailed at the time the sediment was laid down.

The third group of microorganisms to leave abundant fossil skeletons are the *dinoflagellates*. Like diatoms and forams, dinoflagellates are minute single-celled organisms; they are flagellate protozoa, whereas forams are amoeboid protozoa. Also like diatoms and forams, they are an important component of the marine plankton; a few groups are abundant in fresh water as well. And some of them, the ones of interest to micropaleontologists, are "armored"; that is, they have hard, long-lasting exoskeletons of cellulose plates. These plates are sculptured and ornamented to produce extraordinarily baroque

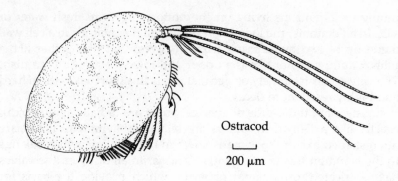

Ostracod

200 μm

patterns, by which different species can be recognized. The example in the drawing is a member of the genus *Ceratium*, which is found in salt water and fresh. Different dinoflagellate species are adapted to waters of different combinations of salinity and temperature; hence their fossils give useful evidence of past environments.

The *ostracods*, the fourth group of organisms that leave plentiful microfossils, are sometimes called *seed shrimps*. An ostracod is a minute crustacean, commonly about one or two millimeters long, whose body is completely enclosed in a hinged, bivalved shell; if it were not for the shrimplike antennae and limbs that protrude through the gap between the valves of the shell (only in living specimens, of course) the animal would look like a miniature clam. There are many species, some living in the sea, others in fresh water. Each species has a unique, recognizable shell pattern.

The tiny, calcified shells are almost indestructible, provided the water surrounding them does not become acidified. Since ostracods thrive in nearly any body of fresh water, it is not surprising that their shells are abundant in the sediments of lakes and ponds. Several thousand species have been discovered, and the ecological requirements of many living species are fairly well known; some can endure only a narrow range of salinities and temperatures. Therefore, if fossils of these species are found at a series of depths in the sediments of a lake floor, one can deduce something of the history of the lake.[2] Nor is that all that fossil ostracods reveal. Some ostracod species live only in ponds or lakes in forested country, others only in ponds or lakes in grassland. They are also affected by the chemistry of the sediment forming the floor of their lake, and hence by the previous history of the area which the lake now occupies. Thus some ostracod species can live only in ponds on sediments that were laid down in the sea, others only in ponds on sediments that were laid down in fresh water. This is because of their sensitivity to the chemistry of the water they inhabit; marine sediments contain more sodium than calcium and more chloride than carbonate, whereas the opposite is true of freshwater sediments.[3]

The fifth group of microfossils consists of the head capsules of the larvae of chironomids (midges). *Chironomids* are the harmless, nonbiting midges to be found "dancing" in aerial swarms over ponds and lakes on calm summer evenings. They lay their eggs in water, and the larvae live in water. In some species the larvae are red blooded and are known to anglers as bloodworms. The bodies of the larvae are soft and maggotlike, but the head capsules (exoskeletons of the head) are hard and resist decay; these are the remains that form microfossils. During its growth, a larva moults four times, depositing four separate head capsules onto the steadily accumulating sediment of its lake. As with other microfossils, so with chironomids: the presences and relative abundances of different species in a sediment sample reveals much about the conditions in which the midges lived.[4]

The sixth group of microfossils we need to consider—the *gyrogonites*—are the fossil "fruits" of stoneworts. Stoneworts are green freshwater algae; they grow upright, like miniature horsetails, in still

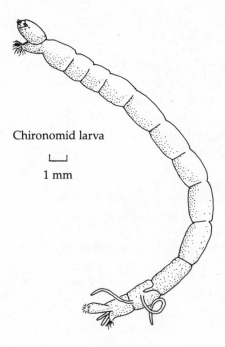

Chironomid larva

⌞__⌟

1 mm

water and secrete a coating of lime that makes them rough to the touch. They are often mistaken for higher plants because, unlike most freshwater algae, they possess what look like stems and leaves. The commonest genera are *Chara* and *Nitella*. The so-called fruit is known as a *nucule*; it is usually about one or two millimeters long. It consists of a fertilized egg (a zygote) tightly enclosed in a cup made of long, tubular cells, twisted in a spiral; the drawing shows a living

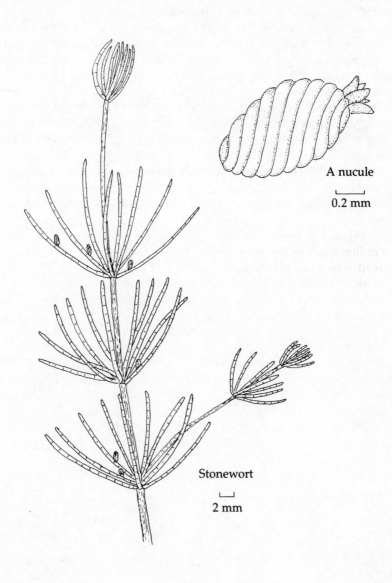

A nucule

0.2 mm

Stonewort

2 mm

stonewort with a fresh nucule (enlarged). Nucules become coated with lime like the rest of the plant, so they readily form fossils. A fossil nucule (a gyrogonite) is a tiny, egg-shaped object with what looks, under the microscope, like a screw thread, which is all that remains of the spirally wound tubular cells that formed the outside of the nucule when it was alive. Like all the other aquatic microfossils, gyrogonites in a sediment give evidence of the conditions that prevailed when they were living.

Pollen

For paleoecologists reconstructing the land vegetation of the past, fossil pollen grains are the most abundant source of evidence. Vast quantities of pollen are produced every year, by many thousands of plant species, and most of it resists decay for an almost unlimited length of time. The task of palynologists is to find and "capture" samples of this dust of microscopically small pollen grains; to examine and identify the several different kinds of grains in each sample; to estimate the proportions of the different kinds; to determine the date (in hundreds or thousands of years B.P.) when the grains were shed; and, finally, to infer the nature of the vegetation that produced them.

Whenever one reads a description of ancient vegetation based on pollen analysis and tries to visualize the landscape as it must have looked when the plants were alive, together with the habitats they provided for animal life, it is useful to recall the long chain of reasoning that led from the raw observations to the final conclusions.

To begin, consider pollen grains themselves. The drawings show a few examples; note the scale bar, which represents ten micrometers (1/100 millimeter). The microphotographs from which the drawings were made[5] are of grains obtained from living plants; therefore, the species of each grain is known with certainty. Identifying fossil pollen grains is not nearly so straightforward; the majority can be identified only to genus, in contrast to macrofossils such as twigs, leaves, and seeds, which can normally be identified to species.

Next, consider how samples of fossil pollen are collected. Pollen from the annual "pollen rain" settles everywhere: on ponds, lakes and bogs, as well as on dry land. What palynologists need are places where the pollen settles out of the air, year after year, in superimposed layers that remain in place, undisturbed, through the millennia. The places that best meet these requirements are lake bottoms and peat bogs. In a lake, the pollen that lands on the surface soon

POLLEN GRAINS

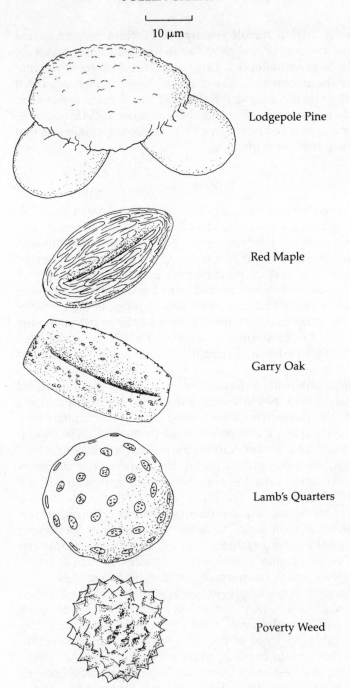

10 μm

Lodgepole Pine

Red Maple

Garry Oak

Lamb's Quarters

Poverty Weed

sinks to the bottom and is incorporated in the sediments that steadily build up on the lake floor; in a bog, the pollen is incorporated in the accumulating peat. From lakes or bogs, researchers take long, vertical cores of material (sediment or peat, respectively); they cut small blocks of this material from the core at a series of depths below the top to give samples of a sequence of ages. They then extract the pollen grains from the material in which they are embedded, mount them on microscope slides, and identify and count them. That, in a nutshell, describes how fossil pollen is collected and studied; subsequent sections provide more details about the process and the problem of determining the age of fossil pollen. But first we delve a bit deeper into the natural history of pollen production and accumulation.

Different plant species differ enormously in the quantities of pollen they produce. Trees and tall shrubs, because of their size, naturally produce more pollen than smaller plants; and for plants of a given size, wind-pollinated species produce far more than insect-pollinated ones. For example, wind-pollinated pines and birches are far more prolific pollen producers than insect pollinated maples and basswood. Likewise for herbaceous plants and low shrubs, the species best represented in the pollen rain are wind-pollinated species such as grasses and sedges. There are also wind-pollinated forbs (non-grasslike herbaceous plants), whose inconspicuous flowers are copious pollen producers. Examples are several members of the goosefoot family and several species of sagebrush, as well as the common agricultural weeds ragweed, common plantain, and various docks and sorrels. Fossils of these "weed" pollens serve as historic markers; a sediment sample containing them in large quantities must have been laid down since the arrival of European settlers in the area where the sample was collected.

Pollen from different plant species also varies tremendously in its buoyancy, hence in the distance it is likely to float through the air before finally settling out on the ground, a bog, or a lake. The pollen of most conifer species is buoyant because each grain has a pair of bladders, as can be seen in the lodgepole pine grain illustrated previously. As a genus, the pines are noted for the abundance and buoyancy of their pollen; it is often found in appreciable quantities a long way from the trees that produced it. Of course the pollen of low-growing plants drifts through the air for much shorter distances than that from tall trees. It has a shorter distance to fall, and the speed of the wind is much less near the ground than at the level of the tree tops.

Therefore, both in pollen quantity and the distance it is likely to travel, there is a marked contrast between trees and tall shrubs on the one hand and herbs and low shrubs on the other. Because of this, pollen from the two sources is usually treated separately when the quantities and proportions found in samples are tallied. More on this in the next section. For the present, it is convenient to note that the two kinds of pollen are labeled AP (for *arboreal pollen*) and NAP (for *nonarboreal pollen*), respectively. The amount of NAP settling out of the air in a forest is negligibly small; the herbs on the forest floor, which are mostly insect-pollinated species, produce only a tiny quan-

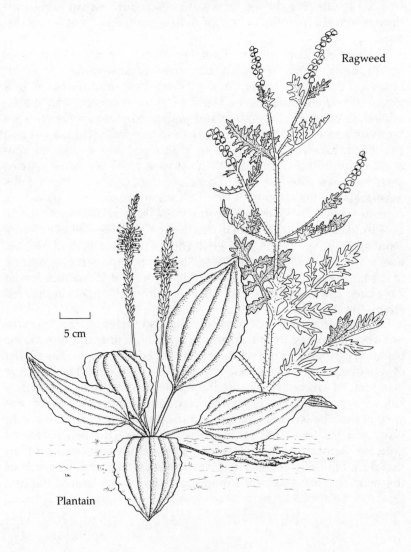

Ragweed

5 cm

Plantain

tity of pollen compared with the trees. Even if the forest is surrounded by grassland, pollen from that source is unlikely to blow into the forest. By contrast, pollen from distant trees is easily blown into grassland, with the result that the pollen rain of grasslands is often "contaminated" with AP; it may form as much as forty percent of the total pollen rain of grasslands.

Now consider a lake and the pollen rain that settles out onto its surface every year. (Most paleoecologists prefer to get their data from lakes rather than bogs, for reasons that will subsequently be made clear). Two questions that immediately arise are, How much pollen is the lake likely to receive? and How large a tract of vegetation will the pollen have come from? First, the question of quantities: the amount is far greater for a lake in forested country than for one in, for example, tundra. Studies of modern pollen rain show that in a typical year, the average number of grains settling out per square centimeter exceeds 6,000 in forest but is fewer than 1,000 in tundra.[6] Second, the area of the source: obviously this depends on many things, chief among them being the area of the lake, the surrounding topography and vegetation, and the force and direction of the prevailing wind in the flowering season. The pollen settling on a lake of "moderate" size (between 10 and 100 hectares in area, say) probably represents the vegetation of an area on the order of 100 to 1,000 square kilometers centered on the lake.[7]

The pollen in a lake sediment is therefore the "average" pollen output of quite a large tract of vegetation. It gives a picture of the commonest, most widespread plant communities. This can be a good point or a bad, depending on the objectives of a particular researcher. It is helpful if one is interested in the prevailing vegetation of a large region. But one must bear in mind that the pollen in a sample may be a mixture derived from a mosaic of separate patches of ground, each with its own distinctive vegetation; an example would be a mosaic of sand dunes and swampy hollows. The contributions of the separate patches would become blended and indistinguishable in the pollen rain settling on a large lake and could give a misleading impression of the makeup of ancient plant communities.

The sediment at the bottom of a lake consists of the remains of dead plants and animals, which disintegrate either in the water or on the bottom after settling, together with wind-blown dust from the land and similar sediments brought in by tributary streams. The number of pollen grains found in a sample of lake mud depends as much on the rate at which lake sediment is accumulating as it does on the amount of pollen rain settling on the lake surface. And (it need hardly be said) sedimentation rates vary widely from one lake to another and also from one century to another in any given lake. In a "typical" lake (if there is such a thing) in forested country, between one and two meters of sediment are laid down on the bottom in a millennium. This is only a very rough figure, of course, and in any case it does not mean that each centimeter, from every depth level, represents an equal time interval of between five and ten years because sediments become more and more compressed as new material accumulates on top.

In any case, a sediment sample of one cubic centimeter (the customary size) would represent several years' worth of pollen rain even if the sediments were completely undisturbed, which they seldom are. In many lakes the topmost twenty or thirty centimeters of sediment are churned by burrowing clams and worms, a phenomenon known as *bioturbation;* also, where the water is shallow, surface choppiness in windy weather stirs up the sediments beneath. Thus the pollen in a sample of lake sediment represents the vegetation of an area over several decades, as well as over many square kilometers. It is this automatic "averaging," over time and over space, that makes lakes a better source of information than peat bogs for most paleoecological work. A lake receives pollen from a large surrounding region, whereas most of the pollen in a peat bog sample comes from plants growing right where the sample was collected; likewise, the pollen in a sample of lake sediment represents a longer time period than the pollen in a sample of peat.

Sediment Cores and Pollen Diagrams

A palynologist's field work consists in collecting cores, either sediment cores from a lake bottom or peat cores from a peat bog. A lake bottom sediment core is obtained by driving a piston corer into the sediment, from a boat in summer or through a hole drilled through the ice covering the lake in winter. Winter has its advantages; it is obviously much easier to locate the site of the cores precisely by taking bearings and measurements to landmarks on the shore, when the surface of a lake is frozen solid. And there are no mosquitoes.

Most cores are five or ten centimeters in diameter. They vary in length. If all goes well (not always, of course) a core extends all the way from the bottom of the water down to the bottom of the sediment laid down during the lifetime of the lake. The surface under the sediment is either bedrock, scraped bare when ice moved over it, or glacial till laid down under the ice. Even if the till consists of clay it is usually easy to recognize, by looking at a core, where the till stops and the lake sediment begins: till is inorganic and pale, whereas lake sediment ordinarily contains plenty of organic material, which .. it dark. It sometimes consists entirely of organic material (dead plankton organisms for the most part) and is then a dark brown, gelatinous mud known as *gyttja*. Depending on circumstances, the thickness of lake sediment—hence the length of a successful core—may be anywhere from less than two meters to more than ten. The quantity of organic material accumulating on a lake bottom depends, of course, on the amount of life the water can support, and this depends, in turn, on the quantity of nutrients dissolved in the water. In a eutrophic lake (that is, one rich in nutrients) the sediment is thick, whereas in an oligotrophic (nutrient-poor) lake it is thin, or even nonexistent. Sometimes sediments are thick because they have been augmented by material brought into a lake by tributary streams. Sometimes they are thin because the lake in which they have collected has not existed for very long, or it may have dried up from time to time and thus not existed uninterruptedly since it was first formed.

When the time comes to examine a core, researchers take samples at regular intervals along it. The usual procedure is to take a sample of one cubic centimeter every five centimeters; sometimes the samples are only half a cubic centimeter in volume, and sometimes they are spaced as close as two centimeters apart. Before they are treated to extract the pollen grains they contain, the samples are merely little blocks of mud. They are first examined to see whether they contain any plant macrofossils, such as seeds or fragments of wood or leaves. Macrofossils can usually be identified to the species level, whereas pollen identification often cannot go below the genus level. (See Interpreting Pollen Diagrams, chapter 3.)

Each sample is then treated with acids[8] to dissolve the material in which the pollen grains are embedded; after considerable processing, a drop of the resulting liquid, which contains the grains, is put on a microscope slide and covered with a coverslip. The grains are identified and counted and the *pollen spectrum* of the sample is calculated; this is a list of the percentages of the pollen grains that belong to each of the recognized types (species, genera, or families) identified in the sample. Two separate pollen spectra are sometimes calculated for each sample, one for the tree or arboreal pollen, and one for the nonarboreal pollen. When this is done, the percentage of spruce grains, for instance, is calculated as a percentage of the total of arboreal pollen grains rather than as a percentage of the grand total of all grains.

To keep the account simple, suppose that the pollen is not separated into arboreal and nonarboreal pollen but that all grains are treated together. Then one pollen spectrum is calculated for each

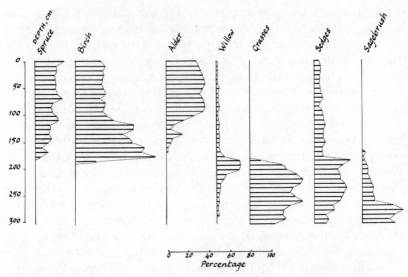

FIGURE 2.1: A percentage pollen diagram[9] showing a set of pollen spectra (each horizontal row of bars is a pollen spectrum) and pollen lines (each irregular dotted line trending from bottom to top of the figure is a pollen line).

sample. The next step in the analysis is to construct a *pollen diagram* for the core being examined. Figure 2.1 shows how this is done. Each horizontal row of bars represents the pollen spectrum of one of the samples taken from one level of the core. The spectra are drawn stacked one above another to show their position in one core; the result is a pollen diagram. Sometimes it is left as is. Sometimes the compiler draws *pollen lines* (the zigzag dotted lines) to show how the pollen percentage of each plant type varied with the passage of time.

So far, so good. But percentage pollen diagrams can be very misleading. Imagine a forest in which almost all the pollen was produced by pine, hemlock, and birch with each species forming thirty-three percent of the total. Suppose the hemlocks were all killed by a severe pest infestation. Then the percentages of pine and birch would each rise to fifty percent, even though the actual numbers of pines and birches in the forest had not changed at all. Because of this defect in percentage pollen diagrams, many palynologists consider that the data used to construct pollen diagrams should always consist of estimates of total quantities of pollen, not merely relative quantities or percentages. The diagrams based on such estimates are known as *pollen influx diagrams*.

Estimating pollen influxes, that is, the number of grains per square centimeter per year, for plants that died several thousands of years ago is, of course, difficult, so difficult that palynologists are divided on the usefulness of the results. Some consider influx diagrams worthless because of the risks of making big errors in estimating pollen influxes, and some consider percentage diagrams worthless because information on actual quantities of pollen is missing from them. As techniques improve, influx diagrams are coming more and more into favor. To construct them, two additional steps are needed besides those required for percentage diagrams. One must discover the actual numbers (rather than the proportions) of grains of each kind in a one-cubic-centimeter mud sample, and one must discover how many years of sedimentation that sample represents.

The first step hinges on the following problem: How do you estimate the total number of grains in one cubic centimeter of mud, recalling that the grains counted were only those on a microscope slide made from a drop or two of the liquid obtained by processing the mud? One way of solving the problem is as follows. The sample is "salted" with a known number of grains of "exotic" pollen of a kind that is easily recognizable and could not possibly occur in the sample naturally, for example, pollen from eucalyptus trees. Or tiny (one-fortieth of a millimeter diameter) spherical plastic beads can be used in place of real pollen grains.[10] It is then assumed that the proportion of the grains in the sample that find their way onto the slide is the same for the naturally occurring grains as it is for the added grains or beads.

The second step (discovering how many years of sedimentation the one cubic centimeter sample represents) requires that the actual age of the sediment be determined at several levels in the core. This is usually done by radiocarbon dating; a brief outline of how the method works is given in the following section.

Figure 2.2 is an example of a complete pollen influx diagram. (Note that in this example the diagram has been blacked in. There is no fixed convention about this; both percentage diagrams and influx diagrams can be blacked in, or can show the individual pollen spectra separately, as in fig. 2.1.) Beside the diagram proper, on the left of the depth scale, the ages of five of the samples, at different levels in the core, are shown; they were found by radiocarbon dating. On the right of the depth scale are the kinds of sediments that make up the core. The ages are given with estimates of their exactness; for instance, the most recent age determined is shown as 6,800 ± 80; this means that the chances are two out of three that the age is in the range of 6,720 to 6,880 years.

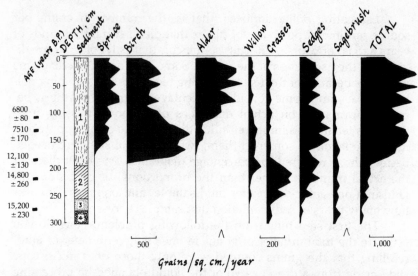

$Grains/sq.\ cm./year$

FIGURE 2.2: Pollen influx diagram for a three-meter sediment core collected from a lake in the Yukon Territory.[11] This influx diagram is constructed from exactly the same data as those used for the percentage pollen diagram in fig. 2.1. Note that the right-hand column in the influx diagram shows total pollen influxes. There is no corresponding column in the percentage diagram since all the percentages sum to 100. The sediments are 1, clay gyttja; 2, silty organic clay; 3, sand; 4, angular pebbles in dense organic clay. Note that three different horizontal scales are used to make the graphs easier to read.

It is easy to see from the figure how the sedimentation rate must have varied over the past 15,000-odd years. Only thirty centimeters or so of sediment accumulated in the (approximately) 700 years between the two uppermost recorded ages, from 6.8k B.P. to 7.5k B.P., but fifty centimeters accumulating in the 400 years from 15.2k B.P. to 14.8k B.P.—and that in spite of the fact that the lower sediments must have been more compressed than those overlying them. Perhaps the sedimentation rate was unusually high around 15,000 years ago because spring runoff chanced to bring extra loads of sand into the lake; the sediment at that level consists of sand.

Dating: The Radiocarbon Method

The age of a sample of organic, carbon-rich material can be determined by measuring its level of radioactivity. The principle of the method, which is known as radiocarbon dating, is as follows.[12]

The element carbon can exist in three different forms, or isotopes, known as carbon 12, carbon 13, and carbon 14; they are usually written as ^{12}C, ^{13}C, and ^{14}C, respectively. The numerals are the atomic numbers of the isotopes, or equivalently the weights of their atomic nuclei. "Ordinary" carbon is not radioactive; it consists almost entirely (ninety-nine percent) of ^{12}C, with an admixture of about 1 percent of ^{13}C. Neither isotope is radioactive. The exceedingly rare ^{14}C, by contrast, is radioactive; its atoms decay, by the loss of an electron, to nitrogen 14 (ordinary nitrogen, with an atomic weight of 14). The half-life of ^{14}C is about 5,730 years; this is the time it takes for half the atoms in *any* quantity of ^{14}C to decay.

The total of all the carbon in the biosphere, the atmosphere, the oceans, and fresh water is known as the earth's *carbon exchange reserve;* it consists of over forty trillion (40,000,000,000,000) metric tons of carbon, of which only sixty tons consists of radioactive ^{14}C; all the rest is nonradioactive, a mixture of ^{12}C and ^{13}C.

Because of its short half-life, ^{14}C would not be present at all if the supply were not continuously renewed; it would rapidly decay until none was left. But it is renewed; cosmic radiation, as it bombards the upper atmosphere, converts a small proportion of atmospheric nitrogen into ^{14}C. Thus the earth continuously gains ^{14}C as a result of cosmic radiation and continuously loses it as a result of radioactive decay, and the net amount in existence at any moment is a mere sixty tons.

Because it is present in the atmosphere, ^{14}C is assimilated by green plants as they photosynthesize carbohydrates. As a result, living plants are all very slightly radioactive; so too, of course, are all other living things as they all, directly or indirectly, depend on green plants for nourishment. Once an organism dies, however, its radioactivity begins to dwindle; it continuously loses ^{14}C because of radioactive decay, and it no longer continuously absorbs it as it did in life. After 5,730 years the remains of an organism will contain only one half as much ^{14}C as the living organism did; after 11,460 years only one-fourth as much, and so on. By comparing the radioactivity of organic remains with that of fresh, living organisms, one can therefore determine how long ago the organism died. This is the principle of radiocarbon dating.

Because of the difficulty of measuring the radioactivity of very weakly radioactive material, the greatest age that can be determined (before recent improvements in technique) is about 40,000 years. Anything older than that is "immeasurably old" so far as radiocarbon dating is concerned, and some other dating technique (there are others) has to be used. For the 20,000-year period that concerns us in this book, however, radiocarbon dating is by far the most important method.

There are innumerable difficulties, needless to say. The chief theoretical difficulty is that the method hinges on the assumption that the rate of formation of ^{14}C in the upper atmosphere has been constant in the past. This is almost certainly not so, but enough is now known about fluctuations in the rate for them to be allowed for.

There are many other problems. For example, the carbon in the specimen to be dated may happen to be contaminated with "older" or "younger" carbon; the older carbon is less radioactive than the carbon belonging to the specimen itself and the younger carbon more radioactive, which leads to overestimation or underestimation, respectively, of the specimen's age. Because of the way in which radioactive material decays more and more slowly with the passage of time, contamination with young carbon leads to much larger errors than contamination with old.

Contamination happens in various ways. A buried specimen can absorb young carbon if the water in overlying fresh humus drains downward onto it; even stored specimens risk being contaminated if they are allowed to become moldy. Old carbon, too can get into a specimen in several ways. A sediment specimen may contain fragments of coal or lignite that were already millions of years old when the sediment was laid down. Also, alkaline lakes often contain old carbon in the form of dissolved bicarbonates derived from the surrounding bedrock. When this happens the aquatic plants growing in the lake acquire old carbon from the water as well as new carbon from the atmosphere, and their fossils therefore seem to be older than they really are.

Still another problem, until quite recently, was that specimens could not be dated by the radiocarbon method unless they weighed at least a few grams. Nowadays, with the new technique of accelerator mass spectrometry, only a few milligrams are needed.[13]

The last point to notice is that, in scientific writing, dates and ages are usually given in radiocarbon years. That is the unit of time used in this book although for brevity the adjective *radiocarbon* has been dropped; dates and ages are given simply in years. However, a radiocarbon year is not quite the same as an ordinary calendar year.[14]

The discrepancy was discovered by checking the radiocarbon ages of small slivers of wood from the interiors of tree trunks against their true ages as determined from counts of annual rings. For this purpose, a long series of datable wood samples was obtained from the trunks of bristlecone pines, using several different trees, both living and fossil. By matching the patterns of the annual rings, it proved possible to assemble a set of trunk sections of overlapping ages that, taken together, spanned more than 8,600 years.

The results showed that the age of a sample as determined by radiocarbon dating is always an underestimate of its true age in calendar years. The cause of the discrepancy is the subject of much geophysical debate, and numerous theories attempt to explain it. Unfortunately, it is impossible to convert radiocarbon years into calendar years unequivocally, except for dates in the comparatively recent past—say the last 8,000 years. It is therefore more straightforward to use radiocarbon years as the time unit in paleoecology. Consistent use of the unit ensures that research carried out in different laboratories can be meaningfully compared.

Dating by Volcanic Ash Layers

In the Cascade Mountains of the Pacific Northwest are a number of volcanoes; some of the best known are Mount Baker, Mount Saint Helens, Mount Rainier, Mount Hood, Glacier Peak, and Mount Mazama in Oregon, whose eruption left the huge crater (a caldera) now occupied by Crater Lake. These and other volcanoes in the region erupted at least once, and some of them more than once, during the past several thousand years, ejecting great clouds of ash that settled out over areas hundreds of kilometers across. Anyone who saw the eruption of Mount Saint Helens on May 18, 1980 (either in real life or on television), can visualize these events.

At every big eruption, some of the ash settled on lakes, sank, and became incorporated in the sediments. The ash is still there, forming distinct layers that can be seen in the sediment cores collected by palynologists. The ash from each eruption, for example, the ash known as Mazama ash from the eruption that formed the Crater Lake caldera, is unique and can be distinguished from the ash ejected in any other eruption. It is recognized by chemical tests and by a microscopic examination of the particles composing it. Therefore, whenever a particular ash layer is found in several different sediment cores, it can be said with certainty that the layer marks the same date in every core; the layer forms a so-called stratigraphic marker.

The age of a volcanic ash layer cannot be judged directly. But sometimes organic remains, datable by the radiocarbon method, are found embedded in the ash; when this happens the ash can be dated very exactly. Other ash layers, in which such embedded remains have not been found, can only be dated by finding organic material above and below them in a core; then the radiocarbon ages of the organic material bracket the age of the ash. A particular ash layer need only be dated in a single core, of course, since its age is the same in all cores, indeed everywhere. Thus, once it has been successfully dated in one core, the layer becomes a date marker in all other cores in which it is found. The most useful ash layers are those that have been accurately dated and that are found over large areas.

Unfortunately, volcanic ash dating can be used only where suitable volcanoes erupted and spread their ash over ice-free lands and waters after the disappearance of the last ice sheets. In northern North America, this means parts of Alaska, southern Alberta, southern British Columbia, and the northwestern United States. More than a dozen dated ash layers are used by palynologists as date markers. Two of the best known are Mazama ash, with an age of 6,720 radiocarbon years, and Glacier Peak B ash (produced by the later of two eruptions of Glacier Peak in Washington), with an age of 11,250 radiocarbon years.[15]

3

Interpreting the Evidence

Some of the Problems

Gathering evidence and interpreting it are two quite different processes. It is one thing to collect and arrange a mass of raw material; it is quite another to interpret the mass and deduce from it a detailed, coherent, internally consistent, chronological "story." As noted in chapter 2, paleoecologists use three kinds of evidence to reconstruct the post-glacial history of glaciated North America: macrofossils, microfossils, and maps of the present-day geographical ranges of a multitude of plant and animal species. Identifying and dating fossils, compiling pollen diagrams (and similar diagrams for other microfossils), and drafting geographical range maps could therefore be called the practical part of paleoecology. The theoretical part consists in inferring from all this evidence what the landscapes of the region looked like as the ice sheets melted, how they changed with the changing climate, and whence and how the immigrant plants and animals came that created and occupied all the different habitats we know today.

Macrofossils provide the clearest and most convincing evidence. For instance, if a 10,000-year-old mastodon tooth is found buried in sediments at a particular place, one can be almost certain that a mastodon died there around 10k B.P. Almost certain, but not quite. It is possible for a deposit of macrofossils to be washed out of a crumbling river bank by a swiftly flowing river and then be carried a long way downstream before being redeposited. When this happens (and it happens to microfossils as well as to macrofossils), the discoverer of the fossils may be misled as to where the original plants or animals lived and died. But misinterpretations from this cause are probably rare.

Reconstructions of ancient vegetation based on pollen diagrams are much less certain. To infer the composition of a region's vegetation many thousands of years ago from a pollen spectrum requires a whole series of deductions, not just one, and each step carries a risk of error. More about this in the following section.

The third kind of evidence used by paleoecologists, the geographical ranges of various plant and animal species at the present day, is the least informative of the three. To infer ancient ranges from modern ranges requires long chains of argument, as we shall see in the discussion on interpreting geographical range maps. Not surprisingly, some of the links in such chains may be rather weak.

When all is said and done, however, and notwithstanding the caveats just mentioned, the post-glacial history of the biosphere in Canada and the northern United States is now fairly well known because it is based on an enormous amount of evidence. Occasionally the evidence is capable of two or more strongly contrasted interpretations, which leads to some spirited controversies, as we shall see in subsequent chapters. More than most sciences, paleoecology depends for its conclusions on indirect inferences; the path from data to conclusions is often long.

But this is not always so. Some deductions are immediate. For example, if a sediment contains fossil fragments of scarabs (dung beetles), one can safely infer that there were mammals living in the area to provide dung.

Scarab Beetle

1 mm

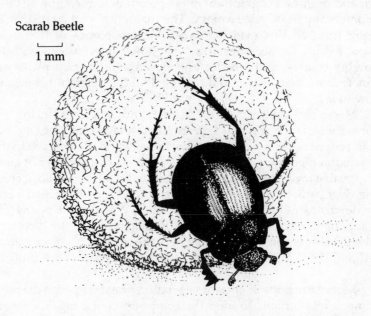

Many other beetle species, plant-eating beetles especially, have very specific food requirements, and when their fossils are found one can confidently infer that their food plants must have been available for them. Indeed food "specialists" of all kinds were (and are) constrained in their migrations; they can only invade an area that has already been invaded by the plants or prey animals they eat. Thus mastodon fossils imply not only the presence of mastodons themselves but also of their favorite food, at any rate in winter, coniferous trees. (Perhaps, like moose, they ate water plants in summer.) Likewise, woolly mammoth fossils imply the presence of tundra plants.

In general, the more specialized an organism's requirements, the later its arrival was likely to be in newly ice-free lands. Parasites can only invade after their hosts. One can sometimes infer the latest date at which a fish species must have reached a particular lake by noting, and dating, the earliest fossil remains of a particular clam species. In many freshwater clams, each tiny newborn larva (a *glochidium*) attaches itself as a parasite to the skin or gills of a fish for the first few weeks of its life. Subsequently, it leaves the host fish, sinks to the bottom, burrows into the mud, and lives independently. But a given species of clam cannot persist in a lake unless its particular species of host fish is present. The fish population can live without the clam population, but not vice versa. Fossil clam shells therefore imply the simultaneous presence in a lake both of the clams themselves and the appropriate fish species; the fish could have arrived long before the clams, but certainly not after them.

Interpreting Pollen Diagrams

A pollen diagram shows how the proportions (or, better still, the absolute quantities) of several pollen species varied over a long stretch of time and, with luck, gives radiocarbon dates for several levels (*horizons*) in the core from which the pollen samples were taken. But the diagram does *not* show how the proportions (or quantities) of the plants that produced the pollen varied, which is what a paleoecologist wishes to know.

Consider a percentage pollen diagram like the one in fig. 2.1. It shows that in the pollen sample taken from a depth of 135 centimeters in a core of lake sediment (with a date of about 7.5k B.P., as can be seen in the corresponding pollen influx diagram in fig. 2.2), the most abundant pollen was that of birch; there was only one-third as much spruce pollen and less still of alder and willow. There was also quite a lot of sedge pollen, but only a negligible amount of pollen from grasses or sagebrush.

What image of vegetation do these figures conjure up? At first thought, one might envisage a rather wet mixed forest made up chiefly of birch, but with considerable spruce, and with alders, willows, and sedges growing in numerous swampy hollows. Such openings as there were would have been wet rather than dry, as suggested by the almost complete absence of grass.

This picture is almost certainly wrong. It is based on three false assumptions. First, that the birch pollen came from tree birch; considering the environment, it is much more likely to have come from a shrubby species of birch. Second, that the quantity of pollen produced by each plant species is proportional to the fraction of the vegetation consisting of that species; equivalently, that the pollen rain "represents" the vegetation that produces it. It quite certainly does not. Third, that the pollen sample obtained from a 7,500-year-old (approximately) sample of sediment is a good representation of the pollen rain at about 7.5k B.P.; this would only be so if every kind of pollen survived prolonged burial in mud equally well. They certainly don't.

The task of interpreting pollen diagrams is therefore fraught with uncertainties. It is worth taking a brief look at the chief difficulties (those mentioned in the previous paragraph) in turn.

The first problem is that of identifying pollen precisely. For example, the pollen grains of junipers and arborvitaes are indistinguishable. In other cases, although it may be easy to identify the genus of a pollen grain, it is difficult to tell the species. Sometimes the grains from different species belonging to the same genus are identical except for slight differences in size. The grains from a single species naturally vary somewhat in size, and if the range of possible sizes for the grains of one species overlaps the range for another species (in the same genus), the identity of a grain of intermediate size is bound to be uncertain. Of course, if the majority of the grains in a sample clearly belong to one of the species, the most reasonable guess, based on probability, is that "uncertain" grains belong to the common species.

Figure 3.1 shows an example of the way grain size varies in two of the pines, lodgepole pine and western white pine. The average length of a lodgepole pine grain (excluding its bladders) is about fifty micrometers (a micrometer is one-thousandth of a millimeter), and of a western white pine grain about fifty-eight micrometers, but there is considerable overlap, and a grain of fifty-five micrometers, say, could belong to either species. The difference between the species' grains

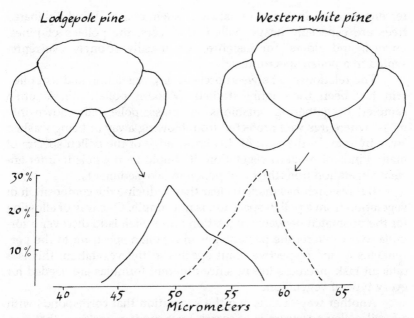

FIGURE 3.1: Size variation in the pollen of two species of pine; lodgepole pine on the left (*solid line*) and western white pine on the right (*dashed line*).[1] Outline drawings of the grains are shown above the graph; they differ only in size. Note that each grain has a pair of bladders.

is statistical rather than absolute, and identifications are probable (often very highly probable) rather than certain.

Some other examples: pollen grains of black spruce are exactly like those of white spruce, except that they are, on average, slightly smaller. In three birch species, dwarf birch with small grains, shrub birch with intermediate grains, and paper birch with large grains, there is so much overlap that the majority of grains could belong to any of the species.[2] In spite of the striking contrast between a tundra of dwarf birch and a forest of paper birch as vegetation types, the pollen of these two birches is hard to tell apart; the pollen of the intermediate species, shrub birch, is even harder to recognize. It is believed that the birch shown in the pollen diagrams in fig. 2.1 and 2.2 is shrub birch.[3]

Once the pollen in a sediment sample has been identified as precisely as possible, and a pollen spectrum drawn, the next problem is that of inferring from the pollen spectrum the composition of the vegetation that produced it. It was noted in chapter 2 that plants differ enormously in the abundance of their pollen. Wind-pollinated trees are especially heavy pollen producers; the pollens of pines, birches, and alders, for instance, are usually strongly overrepresented in a pollen spectrum.

The relationship between present-day vegetation and its pollen rain has been thoroughly studied. Modern pollen rain is often sampled by gathering "cushions" of moss; pollen sifts down into moss, where it is well protected from blowing away or being washed away by rain. In the light of this knowledge of the pollen spectra of many kinds of modern vegetation, it should be possible to infer ancient vegetation from the fossil pollen in lake sediments.

But research has made it clear that deducing the composition of vegetation from a pollen spectrum is not simple. One way of allowing for the mismatch between vegetation and pollen is to discover a formula that converts the percentages in a pollen spectrum to the percentages of the respective plant species in the vegetation; this is a difficult task, needless to say, since different formulas are needed for every type of vegetation.

Another way of inferring the vegetation that corresponds with a fossil pollen spectrum is to search for a modern *analogue*, that is, a modern pollen spectrum that is almost identical to the fossil one. One could then conclude that the vegetation that produced the two spectra must also be almost identical. It has been found, however, that many fossil spectra have no modern analogue. One reason for this is that in populated regions, the modern pollen rain is heavily contaminated with the pollen of agricultural crops. This makes it impossible to obtain a modern pollen sample whose spectrum corresponds with, for example, that of virgin prairie. In many regions, the surviving patches of virgin prairie are too small for their pollen to contribute significantly to the regional pollen rain. Another reason for the lack of modern analogues for many fossil pollen spectra seems to be that ancient vegetation was often utterly unlike anything that exists today. The species were the same—there has not been time since the Wisconsin ice sheets melted for much evolutionary change—but they grew in what we should regard as surprising combinations. More on this topic in subsequent chapters.

Very similar pollen rains can be produced by markedly dissimilar vegetation types. For instance,[4] a fossil pollen sample consisting entirely of nonarboreal pollens like grasses, sedges, and sagebrush suggests either grassland or tundra. It would be easy to distinguish between the two possibilities if the various pollens could be identified to species rather than merely to genus, but usually they cannot. The presence of chenopod pollen is a helpful clue, however, since chenopods are much commoner in prairie than in tundra. Chenopods are members of the family Chenopodiaceae and include such familiar plants as lamb's quarters, Russian thistle, and orache.

Another difficulty with pollen analogues is that it is quite possible for the pollen spectrum of a single tract of vegetation to change over the course of time even though the vegetation doesn't change. Suppose a climatic cooling affects the northern fringe of the boreal forests; even if the cooling is not enough to kill any trees, some species, for example, black spruce, may stop growing cones and producing pollen. Spruce pollen will then be absent from the pollen rain, even though spruce trees are as abundant as they ever were; they can multiply by layering, that is, by the growth of new roots where drooping branches touch the ground. Thus, they will not die out even if they cannot set seed.

Now for the third major difficulty in interpreting pollen diagrams. Not all kinds of pollen preserve equally well. Poplar pollen in particular is seldom well preserved. Thus a sample of fossil pollen

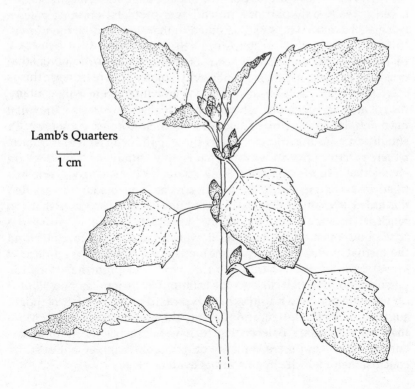

Lamb's Quarters

1 cm

dominated by spruce could have come from a pure spruce forest or a mixed forest of spruce and poplar; there is no way of judging. Larch, too, has fragile pollen that is quickly destroyed by fungi and bacteria.[5]

As a guide to what may have happened to buried pollen over the millennia, experiments have been carried out with modern pollen to discover how fast it decays.[6] The speed of decay depends on both the species of pollen and the type of lake in which it has settled. For example, white pine pollen is very resistant to breakdown, whereas aspen pollen decays quickly. And, as one might expect, pollen deposited in the cold, acid, oxygen-deficient water of peat bogs lasts much longer than that deposited in warm, eutrophic waters where bacteria thrive.

Making allowances for pollen that might have been present in a sediment sample if it hadn't decayed is obviously difficult. Sometimes some of the grains of a particular species are found to be eroded and damaged, in which case it is reasonable to suppose that many other grains have disappeared altogether.

The disappearance of pollen from a sediment because it has decayed is really only part of a much larger problem, one that paleoecologists have become so accustomed to that they seldom mention it. It is that only a small proportion of the plants that once grew in a region, perhaps only five percent of them,[7] have left behind fossil pollen as evidence of their former existence.

This gives rise to a dilemma. When constructing a mental picture of ancient vegetation, should one visualize only those plants that have left incontrovertible evidence of their former presence? Or should one, in imagination, fill in the gaps? For instance, if a fossil pollen spectrum from a lake in what is now Ontario or New England shows that a hardwood forest grew there 5,000 years ago, is one justified in assuming that trilliums, hepaticas, and bloodroot carpeted the forest floor in spring, as they would nowadays? Or is that too fanciful? To stick to the obviously incomplete facts and visualize a hardwood forest without spring flowers is unrealistic. To go beyond the factual evidence and give one's imagination rein is (in a narrow sense) unscientific. Where does the burden of proof lie? With the person who says that absence of trillium fossils implies no trillium? Or with the person who says that a typical trillium "habitat" certainly suggests, even though it cannot prove, the presence of trillium? More than most scientists, paleoecologists have to practice the art of uncertain inference, and there is plenty of room for disagreement over the conclusions to which circumstantial evidence leads.

Interpreting Geographical Range Maps: Animals

The geographical ranges, as they are today, of many plant and animal species provide clues as to where the species came from, and by what route, when the ice sheets melted. The clues are indirect, of course, and conclusions based on them are usually much less certain than those based on fossils. With only the modern range of a species to go on, one can never be entirely sure what its range was at any given time in the past. Even so, plenty of information can be gleaned from modern ranges. They are particularly useful when we attempt to discover the migration routes of aquatic organisms that can spread only

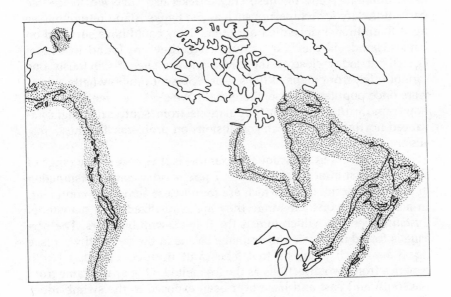

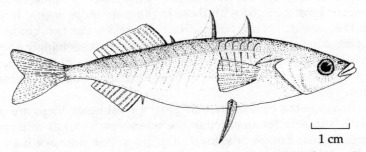

1 cm

FIGURE 3.2: The three-spined stickleback, and its geographical range.[8]

within a connected system of lakes and rivers and are totally unable to get from one drainage basin to another.

As an example, consider the three-spined stickleback, whose present range is shown in fig. 3.2. It is found in both salt water and fresh. The map shows a typical *disjunction*, the existence of two or more regions occupied by the species, separated by wide gaps.

The map immediately suggests that the species must have survived the last glaciation south of the ice on both sides of the continent and must then have made its way northward on the respective coasts as the climate warmed. To assume otherwise would be absurd in the light of this evidence. At the same time, the possibility cannot be ruled out that some of these sticklebacks also survived in ice-free parts of the arctic archipelago such as the Banks Island refugium (see fig. 1.8) and have since died out there. They could have survived on Banks Island, as they can endure cold; they are found in coastal Greenland today close to the Greenland ice sheet.[9] Obviously one cannot infer, from maps of the species' present range, whether there were once populations elsewhere that have since become extinct nor how rapidly the migrating populations from southern refugia advanced northward. To reach conclusions on problems like these, fossils are essential.

Another interesting aquatic example is the present-day range of a parasite that attacks lake trout.[10] There is no apparent disjunction in the range of the trout, which are found clear across the continent. In a small part of their range they are parasitized by a nematode, *Cystidicola farionis*, which infests the fishes' swim bladders. The parasite is found in only a few drainage basins in the far northwest (see fig. 3.3), which suggests that lake trout invaded glaciated North America from two directions as the ice melted. One group came from the south and east and has never been exposed to the swimbladder parasite; another group came from Beringia (see fig. 1.8), bringing the parasite with it. The trout in the northwestern drainage basins can never have mingled with those in other drainage basins; if they had, the parasite would have spread. Therefore the two groups of trout must in fact be disjunct, even though there is seemingly no gap between their ranges. The two ranges are separated only by a line on the ground, namely the height of land separating adjacent drainage basins.

It was noted previously that geographical range maps are particularly valuable for unearthing the migratory histories of aquatic organisms that cannot possibly "jump" from one drainage basin to another. Oddly enough, range maps are also useful in studying the migratory histories of birds, even though birds are the least limited of all organisms in their freedom to come and go as they please.

Here are some examples. As every birder knows, the following three bird species have visibly different eastern and western races;

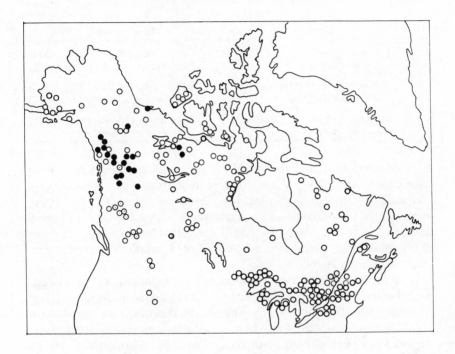

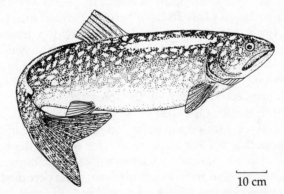

10 cm

FIGURE 3.3: The lake trout and a map showing its range (all dots, both solid and hollow) and that of the parasitic nematode *Cystidicola farionis* that infests it (solid dots).[11]

the difference between the races is so conspicuous that many ornithologists regard them as separate species. The American Ornithologists' Union, which umpires the debate, vacillates; sometimes it lumps each pair of races into a single species, and sometimes it splits them so that the races count as distinct species.

Species Name	Race Names
Common flicker (*Colaptes auratus*)	Eastern race: yellow-shafted flicker Western race: red-shafted flicker
Yellow-rumped warbler (*Dendroica coronata*)	Eastern race: myrtle warbler Western race: Audubon's warbler
Dark-eyed junco (*Junco hyemalis*)	Eastern race: slate-colored junco Western race: Oregon junco

Maps[12] show that the breeding ranges of the members of each pair can be separated by a northwest-to-southeast line across North America like that shown in fig. 3.4. These are not the only bird species divided into eastern and western races with rather similar breeding ranges. It can hardly be coincidence that several pairs of races have matching distributions. A possible, though not proven, explanation[13] is as follows.

It is believed that before the most recent glaciation, the Wisconsin, the species were not differentiated into geographical races. They were spread all across North America in the northern forests that then, as now, spanned the whole continent from east to west. They would have been driven southward when the Wisconsin ice sheets expanded; and while the glaciation was at its height, the ancestors of the modern birds are likely to have become divided into two widely separated groups, an eastern and a western, by the presence in the center of the continent of a cold, sandy, treeless, windswept desert.[14] (For more on this, see Large Mammals and Their Environments South of the Ice Sheets, chapter 5). Woodland birds like flickers, warblers, and juncos could survive in the forests to east and west of the central desert but not in the desert itself. While they were separated, the eastern and western populations had time to evolve into the geographical races that we see now. As the ice melted, the races, now recognizably distinct, migrated northward as inhabitants of the migrating forests; the eastern forests with their fauna occupied land that had been covered by the Laurentide ice sheet, and the western forests with their fauna occupied land that had been occupied by the Cordilleran ice sheet. The breeding ranges of each pair of races meet today along the line shown on the map. The populations mingle along the line, and the two races of each species often hybridize, but even so they have maintained their distinctness.

There are other bird pairs whose modern geographical ranges may indicate their ice age history.[15] The members of these pairs are regarded by nearly all ornithologists as distinct, although closely re-

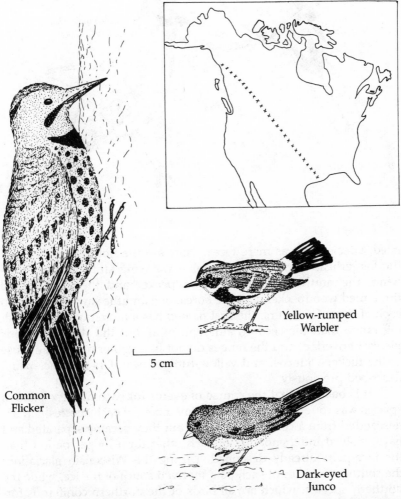

5 cm

Common Flicker

Yellow-rumped Warbler

Dark-eyed Junco

FIGURE 3.4: The common flicker, the yellow-rumped warbler, and the dark-eyed junco. The row of crosses on the map shows, approximately, the line separating the breeding ranges of the eastern and western races of all three species. The eastern races breed in *part* of the area (not all of it) east of the line and correspondingly for the western races.

Northern Shrike

Bohemian Waxwing

1 cm

lated, species, not as mere races. They are the northern shrike and the loggerhead shrike, the Bohemian waxwing and the cedar waxwing, the northern three-toed woodpecker and the black-backed three-toed woodpecker, and the boreal owl and the saw-whet owl. In each of these four pairs, the first named has a more northerly breeding range than the second, although in all but the shrikes there is plenty of overlap, and the ranges do not fit together neatly like those of the flickers, juncos, and yellow-rumped warblers (the race-pairs) described previously.

It is believed that the course of events for each of these pairs of species was roughly as follows. The members of each species-pair are descended from a common ancestor, but they became separated and have evolved independently of each other for a longer period than the race-pairs already described. During the Wisconsin glaciation, the southern species of each pair survived south of the ice, while the northern species, which are all birds of the northern coniferous forests, survived in Asia and Europe, where they still live. Then, as the Wisconsin ice sheets shrank, the southern species of each pair spread northward, while the northern species spread eastward from Siberia into North America by way of Beringia (see fig. 1.8) and now occupy extensive North American ranges.

Interpreting Geographical Range Maps: Plants

Now we come to plants. A large number of plant species with disjunct ranges are known; a particularly common form of disjunction is the existence of one or more small, separate *outliers*, where a species grows in a circumscribed patch of land far removed from its main range. An example is white dryad, a plant of arctic and subarctic tundra and the alpine tundra of the western mountains. It also grows on the north shore of Lake Superior, in a small, isolated, southerly outlier of the arctic tundra surrounded by forest.[16]

There are three possible explanations for such outliers in a plant's range, and they are pictured diagrammatically in fig. 3.5. The first possibility (map A in the figure) is that, when ice sheets covered most of the area, the outlier happened not to be covered; it was either a nunatak or a coastal refugium (see chapter 1). In either case it was an "island" surrounded by ice throughout the Wisconsin glaciation, where the species managed to survive; it is also assumed that the refugium population has not expanded appreciably since the ice disappeared but has continued as a disjunct outlier until the present day. The present range of the species is shown in the map at the bottom of the figure.

The second explanation (map B) is that the range of the species was at one time continuous but that, because of environmental change, the species has since disappeared from much of its earlier range, leaving the outlier separated from the rest of the range. The population remaining in the outlier is known as a *relict* population,

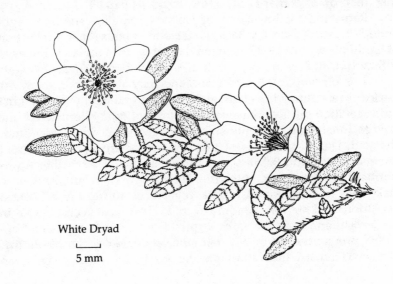

White Dryad

⊢——⊣
5 mm

one that has been left in possession of a piece of ground even though the main range of the species has contracted or shifted away. This is the most likely explanation for the disjunction in the range of white dryad. While the ice sheets were melting, there was probably a strip of tundra immediately south of the ice, and forest south of the tundra. As the ice front receded northward, the vegetation zones migrated northward in its wake. For a while, the north shore of Lake Superior was part of the tundra zone; then the advancing forest invaded, but a relict island of tundra vegetation survived where the soil happened to be too shallow and rocky for trees. The outlier was surrounded but not overrun by forest. There are other tundra species besides white dryad in the outlier, as one would expect if this explanation is correct.

The third explanation for outliers (map C) is that seeds of the plant species concerned happen to have been carried a long way by the wind, and have then landed, germinated, and become established far from their main range. The process is known as *jump dispersal*. It is certainly a possible reason for outliers, but in most cases it seems rather unlikely. The first two explanations are believed by most paleoecologists to account for the majority of the plant disjunctions in our area. In a sense they are the same explanation: the process illustrated in map A is only a particular example of that in map B, with the formation of an ice sheet being the environmental change that destroyed the plant over part of its earlier range.

There is no way of telling when an outlier became established. It may date back more than 20,000 years (map A), somewhat less than that (map B), or perhaps only a few decades (map C).

Returning to the matter of isolated refugia, especially high-altitude nunataks: it may be reasonable to suppose that a collection of hardy plants survived in sheltered crannies on a nunatak, but even if they did, it does not automatically follow that their descendants would be present at the same spot, and only at that spot, today. As the ice sheets thinned, and nunatak islands expanded, why shouldn't the vegetation on them have spread outward too? Alternatively, the ice age inhabitants of a nunatak may have died off as conditions changed. During the warmest part (the hypsithermal) of our present interglacial, around 8,000 years ago (see chapter 1), the climate was warmer than it is now and the tree line higher. Therefore many nunataks that had supported tundra vegetation during the Wisconsin glaciation, and are now tundra covered, must have been forested in the hypsithermal; as the forests spread up the mountains and engulfed previously tundra-covered summits, most of the plants that had survived on them through the glaciation would have been

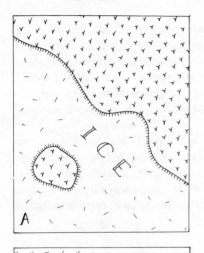

A

B

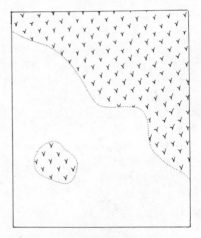

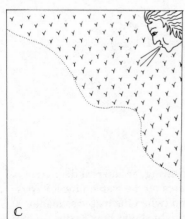

C

FIGURE 3.5: Three theoretically possible precursors (left column) of the present-day range map shown at right. **A,** The species survived on a nunatak and has persisted in the same place without spreading. **B,** The species was formerly widespread but has since died out over part of its earlier range. **C,** The outlier was established by jump-dispersal (perhaps by wind-blown seeds) from the former continuous range of the species.

crowded out by species adapted to warmer conditions. In a manner of speaking it was warmth, rather than cold, that killed off the majority of the ice age survivors. The warmth enabled vigorously competitive plants to exclude the cold-tolerant species. This may explain why nunatak survivors are not more common than they are, although there certainly seem to be some, as we shall see in subsequent chapters.

One probable nunatak survivor is the arctic and subarctic plant pale corydalis, whose modern range is shown in the two maps in fig. 3.6. The maps themselves illustrate a problem that always arises when range maps are drawn. Consider the three ways in which such maps could be drawn. The first, and best, has already been illustrated

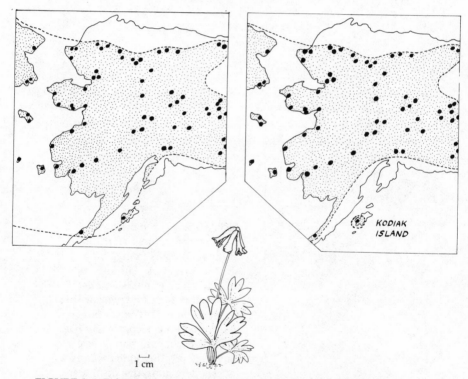

1 cm

FIGURE 3.6: Pale corydalis and two maps showing, at different degrees of detail, its geographical range. The stippled area on the map on the left covers all sites where the species had been found (when the map was made) disregarding possible gaps. The map on the right shows how Kodiak Island may have served as an isolated glacial refugium for the species.

in fig. 3.3. On the map in that figure, every dot (both solid and hollow) shows a point where the species concerned (a nematode parasitic on trout) was searched for; solid dots show where it was found, hollow dots where it was not found.

The other two possible ways of drawing range maps are illustrated in both the maps in fig. 3.6. One can place a dot on the map to show every point on the ground at which the species concerned has been collected or one can outline what are believed to be the limits of the species' range and shade the area within. The "dots" method is obviously the better; it takes nothing for granted. But it leaves an obvious question unanswered: Where gaps occur among the dots, does it mean that the species does not live there, or merely that nobody has looked for it there? A map like that in fig. 3.3 does answer this question, but maps showing both "presences" (solid dots) and "absences" (hollow dots) are not feasible for organisms such as a plant species. One cannot draw up an exhaustive list of the places where a plant was *not* found in the way that one can list the places where a trout was caught, dissected, and found not to contain a parasite.

The shaded area method of drawing range maps has obvious drawbacks. First, the meaning of the shaded areas is unclear; the species does not carpet the ground everywhere within them. Second, the justification for the boundary lines is usually rather vague; if one of the dots on the dot map is separate from the rest, should it be enclosed within an all-inclusive range outline as in the map on the left in fig. 3.6? Or should it be set apart as a separate outlier as on the right? It depends entirely on the judgment of the person drawing the map. The maps in fig. 3.6 were both drawn by the late Eric Hultén, a leading authority of his time on arctic botany.[17] A range map intended for printing is always a simplified version of an original dot map drawn on a much larger scale, and some of the detail in the original has to be sacrificed when the scale is reduced. How the simplification should be made depends on the purpose of the map makers. The map on the right illustrates an account of the paleoecology of Kodiak Island, a well-studied glacial refugium; in the context it is natural to show the Kodiak Island population of pale corydalis as an outlier, separate from the rest of the range. The map on the left (adapted from Hultén's *Flora of Alaska*) is more simplified; the outline of the species' range is drawn so as to envelope all the sites where it has so far been found because, in the context of a flora, no one site merited special attention.

This rather lengthy discussion of range maps has been necessary because of their importance in discovering and describing the history of ecological change. Notice that maps have two functions: discovery and description. A map may be part of the evidence leading to new hypotheses; it is also the best way of displaying those hypotheses. But a map, by itself, can never prove a hypothesis conclusively.

4

The Migration of Vegetation

Shifting Zones of Vegetation

Most people are familiar with maps of the major vegetation types of North America, showing a strip of tundra in the far north, a strip of coniferous forest immediately south of the tundra, and so on. It is easy to envisage similar maps showing the vegetation zones as they must have been at different times in the past, when great ice sheets covered much of the north. And it is not difficult to visualize a motion picture version, with the zones creeping southward as the ice sheets grew and then creeping northward again when the climate warmed and the ice sheets slowly melted. It is so easy to visualize such scenarios that one is apt to forget that the southward shift of the vegetation zones as the ice expanded, and their northward shift as the ice contracted, were radically different phenomena. Obviously, vegetation does not "creep" in the way that strips of color on a motion picture of a map do; likewise plants cannot "migrate" or "advance" or "retreat" in the way that animals can. To use these words in connection with plants is to speak in metaphors that may be useful as shorthand but are apt to disguise the true nature of what they represent.

The responses of vegetation to growing ice sheets on the one hand, and to shrinking ice sheets on the other, were entirely different. When an ice sheet expanded one of two things happened, depending on the cause of the expansion. If the ice spread because of climatic cooling, then the cooling would also have affected the vegetation ahead of the ice front. The less hardy plants gradually died off, and permafrost (perennially frozen ground, see following section) formed, seriously inhibiting the growth of trees. In other words, a strip of tundra would have developed or, if one was already there, it

Ice surging over forest

would have widened, ahead of the ice. But if surging ice lobes caused the expansion (see chapter 1), full-grown forests would have been overrun by the ice, which crushed and buried them.

In either case, plants did not and do not "migrate." What does migrate is only an abstraction, a line on a map representing the margin of a vegetation zone. The plants themselves die.

Zone margins also migrate, of course, when an ice sheet shrinks. What actually happens is that seeds are blown into the area laid bare by the melting ice, where they germinate and grow. The kinds of seeds blown in depend on what plants are present in the periglacial zone (the unglaciated region adjacent to the ice). And the success of the seedlings in establishing themselves depends on the conditions in the newly ice-free land. But whatever the details, the vegetation renews itself on land that may have been ice covered for tens of millennia.

How the renewal of vegetation takes place is the topic of this chapter. For paleoecologists, vegetation has two aspects: It is part of the living world that is studied for its own sake, and it is also the biological setting that, combined with the physical setting discussed in chapter 1, forms the total environment of all life. Before embarking (in chapter 5) on a chronological history of how life took over in glaciated North America as the ice disappeared, it is therefore desirable to discuss a few general questions such as: How does vegetation first get started on a completely inorganic substrate? Do different plant species advance independently of one another or as integrated communities? How fast is the advance? The answers to these questions or, at least, opinions about them are implicit in all reconstructions of our ecological history, so it is worth tackling them at the outset.

First, however, we must inquire into the starting conditions: What was the vegetation like around the periphery of the ice sheets at the time of glacial maximum? It is a subject on which there is much disagreement among experts.

The Starting Conditions

First consider the region northwest of the ice sheets, that is, unglaciated Beringia. Presumably it was in the grip of permafrost, and consequently devoid of trees, except in a few unusually mild spots (see Beringia and Its Big Game, chapter 7). Wherever permafrost prevails, the soil below a certain depth remains frozen all through the year; it never warms sufficiently to thaw. The layer of soil above the perennially frozen layer, which does thaw out for a short period each summer, is known as the *active layer*. The kind of vegetation that can grow on the active layer depends on its thickness. If it is thin, say less than fifty centimeters, it cannot support full-grown trees. The frozen ground below the active layer blocks the downward growth of tree roots and also prevents summer meltwater from draining away. The layer therefore becomes sodden with water that is cold, deficient in oxygen, and lacking in nutrients, this last because the low temperature slows the decomposition of dead plant remains. The occasional groups of stunted trees that do manage to establish themselves often form a "drunken forest"; their trunks lean in all directions because frost-heaving takes place and the rising mounds of freezing soil tilt the trees growing on them. Therefore if, as seems certain, most of Beringia was covered by permafrost, it must have been mostly treeless. It may have been covered by tundra vegetation much like the modern arctic tundra, but this is a topic of considerable disagreement and we return to it in chapter 7.

The state of affairs south of the great ice sheets is also a topic for debate. There was certainly some tundra between the ice front and the extensive evergreen forests that stretched across most of the continent farther south, but it seems unlikely that tundra formed a continuous, uninterrupted strip. There is evidence that, in places, there was no gap at all between the ice front and evergreen forests. Probably there was a "vegetation mosaic," with patches of evergreen forest alternating with patches of treeless tundra; this is suggested by the odd mixtures of species whose fossils have sometimes been found together. For example,[1] fossils of forest trees (spruce and pine) have been found mixed with fossils of tundra plants (bog blueberry, dwarf willow, and white dryad) in deposits in Massachusetts.

Presumably when they were alive, the plants did not grow mingled together as members of a single vegetation type. The tundra plants are intolerant of shade and are not likely to have grown on the floor of an evergreen forest. It therefore seems more likely that there were separate, contrasted patches of vegetation, perhaps with forest growing where the soil was relatively deep, and tundra on thinner soil.

It is not known how closely the tundra of ice age times resembled modern tundra. The tundra of 18k B.P. in what is now the northern United States can hardly have been indentical with the modern tundra of arctic and subarctic Canada. Because of the lower latitude, summers may have been appreciably warmer, although there is also evidence for arctic cold.

Some of the evidence for extreme cold and, in places, permafrost south of the ice sheets is geological.[2] Landforms implying the past existence of permafrost are found today in places known to have been near the ice margin, for example, in Washington, Montana, Wyoming, North Dakota, Wisconsin, Iowa, and New Jersey. Casts of what were once wedge-shaped blocks of ice in the soil have been found in these places; these are, in fact, fossil ice wedges and are a sure sign of past permafrost. Patterned ground is another sign of extreme cold; repeated freezing and thawing of the soil near the surface causes it to become sorted into a fairly regular pattern of raised polygons outlined by a network of rock-filled ditches. So-called fossil patterned ground has been found in several places south of the southernmost ice limit, for example, in Idaho and Pennsylvania. The pattern has chanced to persist even though the processes that originally caused it to form no longer operate. The growth of forest, or erosion by streams, destroys the pattern of patterned ground; probably only a little now remains of what were once far more extensive tracts.

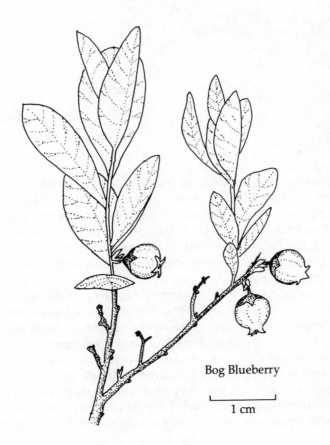

Bog Blueberry

1 cm

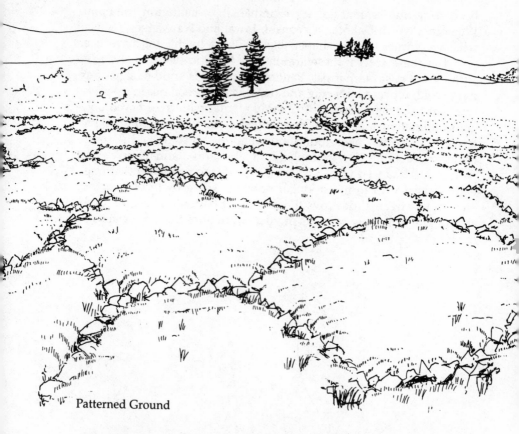

Patterned Ground

There is also compelling biological evidence for cold, tundralike conditions south of the ice sheets. Pollen and larger fossils of such tundra plants as dryad, crowberry, and moss campion have been found,[3] as have fossils of tundra animals such as lemmings, arctic ground squirrels, caribou, and tundra muskoxen.[4] By far the most spectacular of the tundra mammals were woolly mammoths, which may have been plentiful.[5]

The presence of woolly mammoth fossils in a region is usually taken to mean that the region was tundra covered when the mammoths were living. Woolly mammoths appear to have been well adapted to a cold climate.[6] They had smaller extremities (ears, feet, tails, and trunks) than other species of mammoths; their thick hides were furred with long guard hairs over a soft, woolly undercoat even on trunk, ears, and tail, and they were insulated against the cold by layers of subcutaneous fat. Their huge molar teeth (see fig. 8.2) were faced with numerous hard, close-set ridges, which presumably enabled them to chew coarse tundra vegetation. And the pattern of wear and tear on fossil tusks suggests that the mammoths used their tusks as scrapers to clear the snow from their food plants.

The evidence is therefore strong that a strip (but not an uninterrupted strip) of open ground with low vegetation lay immediately south of the great ice sheets at the time of glacial maximum; the low vegetation may have resembled the modern arctic tundra in general appearance, or it may have been more like a grassy steppe. Even if it was tundralike, it almost certainly differed in detail from modern tundra.

Conditions in the Newly Deglaciated Land

As the great ice sheets began to shrink, expanding areas of open ground became available for plants to grow on. Until at least some vegetation existed, there was no habitat for animal immigrants. It is

Woolly Mammoth

20 cm

interesting to consider how plants got started on the lifeless terrain left uncovered as the ice disappeared. What was the terrain like, and what kinds of plants invaded?

The plant pioneers, those that established themselves first, must have grown from seeds that were carried in by the wind from the periglacial vegetation onto the newly available land. Most of these seeds would therefore have come from tundra plants; but at the southern edge of the ice, where the tundra strip was discontinuous and probably nowhere more than 100 kilometers wide,[7] coniferous forest was close at hand and abundant tree seeds, especially of spruce, must have been available. They could have been carried long distances as the winds were strong; a persistent anticyclone over the ice caused frequent gales, and there were no trees to act as brakes or to screen out the wind-borne seeds.

The species of seeds available would have been one of the chief factors controlling the nature of the first plant cover on newly ice-free land, but it was certainly not the only factor. The climate and the condition of the ground surface were equally important. As to climate, two scenarios can be envisaged. In theory, the shrinkage of the ice sheets could have been caused solely by diminished snowfall over the interior of the ice-covered regions and consequent starvation of the ice (see chapter 1); there need not have been any warming of the climate. If that had happened, the newly exposed ground would quickly have become perennially frozen (if it wasn't already) and there would have been no ice front lakes. However, there were ice front lakes—enormous ones—south of the waning ice sheets (see figs. 1.3 and 1.4). It follows that the ice starvation scenario is untenable, at any rate in the south, and that the disappearance of the ice must have been the outcome of melting, brought on by a warming of the climate.

The kind of vegetation that could invade the newly available ground around and between the lakes depended on whether the ground was frozen. It need not have been. If the ice, when it advanced, had surged over unfrozen ground, the latter would have been insulated by the overlying ice and could have remained unfrozen even though the climate cooled sufficiently to freeze the uncovered ground ahead of the ice margin.[8] The first vegetation to appear was governed by the state of the ground. On unfrozen ground, forest could develop rapidly, but where permafrost prevailed, forests could not develop until the land was free of it. Such a state might be reached soon after the ice disappeared or not for thousands of years. The presence or absence of permafrost controlled the speed at which forests shifted north in the wake of the melting ice. Until the ground was thawed, it could only support tundra.

The length of time required for vegetation of any kind to get established on ground laid bare by the melting ice must have been governed by the time it took for soil to develop. Soil formation in situ, starting from scratch on a newly exposed inorganic surface, is usually slow. On newly deglaciated land, the process was sometimes speeded up by the addition of ready-made soil blown in on the wind from nearby unglaciated areas. The first layers of wind-borne soil provided a seedbed, possibly a fertile one, for the first wind-borne seeds. The resultant plants then contributed to further soil development in two ways: first as drift fences causing more wind-borne soil to accumulate and then, when they died, as compost that enriched the soil still more. Obviously, the stronger the winds and the nearer the source of dry, unconsolidated soil, the faster the process of soil development.

The texture of the newly ice-free surface also affected soil formation. Drifting soil would have been blown away as easily as it was blown in if the surface consisted of smooth bedrock, polished by moving ice. It would have been able to accumulate much faster where the bedrock was covered with a rough coating of glacial drift. The type of drift is important, too. If it consisted of till (material laid down directly by the ice), which normally contains an abundance of clay and fine silt, it would have been capable of holding large reserves of water. If it consisted of outwash (material deposited by meltwater rivers), it would have had little water-holding capacity. Most outwash consists of a mixture of coarse sand, gravel, pebbles, and larger rocks, and water quickly drains away through it.

Where ready-made soil was thin or nonexistent, the development of vegetation must have resembled what is happening at the present day in places such as Glacier Bay, Alaska, where newly exposed till, uncovered by a retreating glacier, is in the process of being colonized by plants.[9] The first invaders, willows and cottonwoods, show obvious signs of a lack of nitrogen; they grow prostrate on the ground and their leaves are yellow. Things cannot improve until after the establishment of nitrogen-fixing plants. These are plants whose roots are inhabited by bacteria able to take in gaseous nitrogen from air spaces in the soil and convert it into soluble nitrogen compounds that the plants can absorb and use. Such nitrogen-fixing plants (or nitrogen fixers) are the only plants able to flourish on a soil deficient in nitrogen; when they die, their decomposed remains amount to nitrogen fertilizer. The best nitrogen fixers native to the Glacier Bay region are yellow dryad and Sitka alder. Extensive mats of the former and dense thickets of the latter quickly develop and enrich the soil.

If trees are to invade successfully, another requirement is the presence of fungi that form *mycorrhizae* with the tree roots. A mycorrhiza is a tree root with filaments of fungus (of an appropriate species) growing for some distance around it, touching it, and even inside it in the intercellular spaces. The fungal filaments enable the tree to absorb nutrient-rich soil water from a far greater volume of soil than would be accessible to roots without the fungus. Most forest trees depend on mycorrhizae for healthy growth and do poorly without them. Therefore, the development of a soil in which trees can thrive must wait until the spores of mycorrhizal fungi have been blown in.

Newly deglaciated land may often have consisted of a patchwork of till and outwash, with contrasted vegetation on the two surfaces; much richer vegetation would have developed on till than on outwash because of till's better moisture-retaining capacity. The contrast persists to this day in parts of Michigan, where a luxuriant mixed forest of sugar maple, yellow birch, eastern hemlock, and eastern white pine has grown up on till but where only jack pine, which does well on dry, sandy soil, has established itself on outwash. The pattern of vegetation as it appears on a map, which coincides with the inorganic pattern of till and outwash, is believed to have existed continuously ever since the ice disappeared[10]; however, the mixture of tree species making up the moist forest must have gradually changed during the course of time, as we shall see later.

Recall (chapter 1) that in some places new ice surged over partly melted old ice and deposited a layer of till on top of the old ice. When melting resumed, only the upper, younger ice layer responded quickly to the warming trend; the lower, older layer was insulated by the overlying till and remained as buried stagnant ice that melted very slowly indeed. In some places it may have taken as much as 3,000 years to disappear completely; however, its presence, deep underground, had no effect on events at the surface, where normal vegetation developed.

Besides denuded bedrock and glacial drift, one other initially lifeless habitat awaited colonization when our interglacial began. This was the fresh water of the numerous proglacial lakes that rapidly formed along the receding ice front (see chapters 1 and 9). Aquatic organisms of all kinds must have invaded them by degrees. Animals unable to endure drying at any stage in their life cycles—fishes, for example—could have reached the lakes only through rivers draining into them. The richness of the freshwater ecosystems that developed in each lake would thus have depended on the richness of the refugium (or refugia; there could have been several) from which the lake received its immigrants and also on how long the lake persisted before it drained away. A newly exposed lake bed would then be left ready for colonization by terrestrial plants.

The Invasion by Plants

When seeds were blown in from the south onto the variety of newly available surfaces left bare by the melting ice, and when they succeeded in germinating and becoming established, the result was the beginnings of new vegetation, sparse at first but gradually becoming denser and finally culminating in forests. The first forests were probably quite unlike modern forests. Obviously, it was not forests as such that colonized the ice-free land but individual trees; the kinds of forests that these trees formed, and the way in which they subsequently changed, depended on the order of arrival of the different tree species.

For example, consider the forests now growing in the region of the present-day Great Lakes. They consist of a mixture of evergreens and hardwoods; the commonest evergreens are eastern white pine and eastern hemlock; the commonest hardwoods are sugar maple, beech, yellow birch, basswood, white and red ash, and white and red oak. A wealth of other species are present besides, making these forests by far the most diversified of any now growing in a once glaciated area.

This wide array of species did not all arrive at their present locations at the same time, however. They spread northward from their different ice age refugia, at different starting times, and with different speeds. The migratory behavior of many important tree species has been unearthed by Margaret Davis, one of North America's leading palynologists. She has pieced together the results of studies on more than sixty sediment cores.[11] The arrival of a particular species at a particular site naturally causes a sudden increase in the quantity of that species's pollen. The time of arrival can therefore be discovered by determining the radiocarbon age of the sediment sample that records the sudden increase.

It turns out that each species has a unique migratory history. Take the four species illustrated in figure 4.1 as examples. It is believed that eastern white pine and eastern hemlock survived the Wisconsin glaciation in a refugium in the eastern foothills of the Appalachians and on the adjacent coastal plain. In contrast, chestnut, of which only a few specimens now remain because of a devastating epidemic of blight earlier this century, and maples are believed to have survived in a refugium around the mouth of the Mississippi River. The two pairs of species therefore began their northward spread from entirely different starting points.

They did not all begin their spread simultaneously. For instance, eastern hemlock began its northward migration about 1,000 years later than eastern white pine, presumably for ecological reasons. Eastern white pine can succeed on poor, dry soils, exposed to full daylight, but eastern hemlock requires a richer, moister soil and the shade of other trees. Therefore, conditions did not become suitable for the advance of hemlock until generations of plants had lived, died, decayed, and enriched the soil, and a forest of other trees had developed to provide shade.

The separate species migrated at different rates, too. Davis found that for the four trees we are considering here, the average rates of advance were as follows: eastern white pine, 300 to 350 meters per year; eastern hemlock, 200 to 250 meters per year; maple, 200 meters per year; and chestnut 100 meters per year. These are long-term averages, of course; the advances probably took the form of sudden spurts separated by long pauses. Of this particular group of trees, those with winged seeds (the pines, hemlocks, and maples) all migrated faster than chestnut, which depends on animals to transport its heavy, unwinged seeds. But it is not invariably true that wind-dispersed trees are the fastest migrants; oaks migrated at an average rate of 350 meters per year, in spite of their heavy, animal-dispersed acorns.

Even so, lightweight wind-borne seeds certainly gave their possessors an advantage. It is likely, too, that the rate of advance of a species differed in different regions, depending on the force of the prevailing winds. This seems to have been true of white spruce. In the eastern half of the continent, its progress from the latitude of Pennsylvania to the coast of Labrador took about 7,000 years (from 14k B.P. to 7k B.P.), and it covered the distance at a steady pace. Pollen diagrams from eleven lakes ranged along the route show how spruce arrived at these sites one after another at fairly regular intervals, advancing at about 300 meters per year on average[12] (fig. 4.2).

The course of events in the west was quite different, however. After steadily advancing northward from the vast evergreen forests that covered what are now the Great Plains, white spruce made a sudden northward "sprint" of 2,000 kilometers to reach the shore of the Beaufort Sea. Over this last section of its route its migration was perhaps ten times as fast as in the east. The sudden increases in spruce pollen recorded at ten lakes spread along the route of the

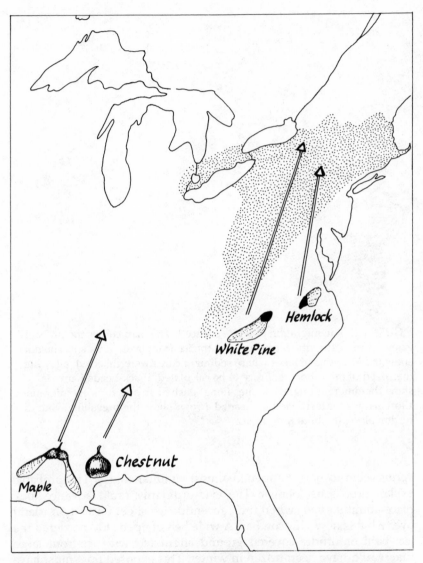

FIGURE 4.1: The different refugia, and different migration speeds, of four species of tree. Maple and chestnut spread north from an area around the mouth of the Mississippi; eastern white pine and eastern hemlock, from the east coast. The lengths of the arrows are proportional to their speeds of advance. The shading shows the area in which the ranges of all four species overlap at the present day.[11]

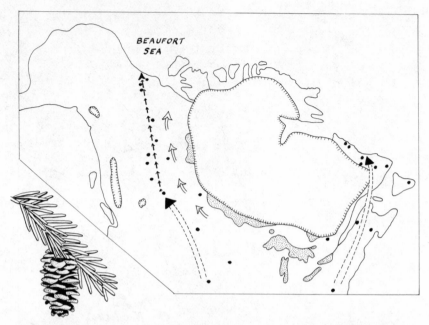

FIGURE 4.2: The migration of white spruce.[12] The map shows the probable extent of the Laurentide ice sheet and the ice front lakes at 9k B.P. The dots show the locations of lakes where sediment cores were collected, allowing the speed of migration of spruce to be calculated. Black-headed arrows show the different migration rates (long, dashed arrows for slow migration; short arrows for fast). Hollow-headed arrows show the prevailing wind direction along the border of the ice.

sprint seem to have happened almost simultaneously. The probable explanation is as follows. The extraordinarily rapid advance took place about 9k B.P., when the Laurentide ice sheet formed an island over what is now Hudson Bay. A wide belt of open land bordered the ice; bare or tundra-covered ground alternated with ice-front lakes that would have been frozen in winter. This exposed tract must have been constantly swept by fierce southeasterly winds blowing clockwise around a persistent anticyclone over the ice; the strong winds, blowing over treeless land, brought about unusually rapid dispersal of the winged seeds of white spruce.

Another species whose migration has been closely studied is lodgepole pine. The present-day range of one variety of this species (*Pinus contorta* var. *latifolia*) is shown in figure 4.3. It was thought at one time that the species had survived the Wisconsin glaciation both south of the ice sheets and north of them, in Beringia, since it is found today in part of what was unglaciated Beringia. Studies of pollen diagrams from a long north-south row of sites show,[13] however, that lodgepole pine has reached its present northern limit very recently, possibly less than a century ago[14]; it has advanced through a distance of 2,200 kilometers in about 12,000 years (that is, at an average speed of a bit less than 200 meters per year), and its range is

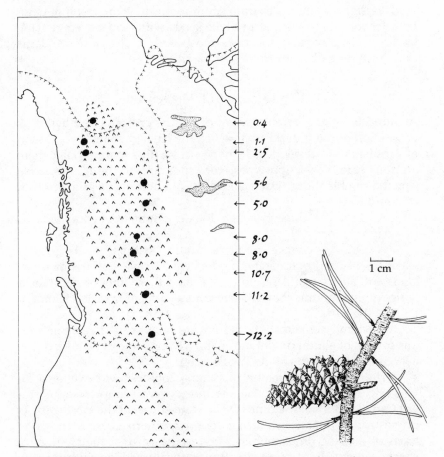

FIGURE 4.3: The migration of lodgepole pine. The map shows the modern range of the inland variety of the species, the limits of the ice at the time of maximum glaciation, and the locations (black dots) of lakes where sediment cores were collected. The numbers on the right, opposite each dot, show the time, in thousands of years B.P., when lodgepole pine is believed to have reached the site.[13]

probably still expanding northward. While this long, slow advance was in progress, a gradual genetic change was taking place in the trees in the vanguard. Two Canadian palynologists, Les Cwynar and Glen MacDonald, found that compared with trees growing in the south, those in the north produced seeds with smaller wing loadings. The wing loading of a winged seed is the ratio of the weight of the seed to the area of the wing; clearly, the smaller the value of this ratio, that is, the lighter the seed relative to its wing size, the farther it will be carried by the wind before settling to the ground. The change in seed "design," as one goes from south to north, is the result of natural selection; the trees that produce seeds with below-average wing loadings are obviously the ones most likely to leave descendants growing a long distance from their parents.

The Renewal of Vegetation

As multitudes of different species of plants, one after another, took possession of the ground from which the ice had melted, vegetation of various kinds developed. In northern North America today, there are four chief vegetation types, corresponding to the four possible climatic combinations: warm and dry; cold and dry; warm and wet; cold and wet.

A warm, dry climate leads to the development of grasslands, as in today's prairies.

A cold, dry climate produces tundra. At present, the greatest expanses of tundra are arctic tundra, which covers all the arctic islands and all the mainland north of the northern limit of trees; there is also alpine tundra, above the elevational limit of trees in the mountains of the west.

A warm, wet climate favors forests such as the largely deciduous forests of southern Ontario, Pennsylvania, and New York or the evergreen rain forests of the Pacific coast.

The fourth combination of factors, coldness with wetness, favors taiga or boreal forest, the enormous evergreen forests of northern Canada south of the tundra. The wettest part of the taiga consists of *muskeg*, the name given to forested northern bogs. Many large tracts of muskeg probably developed in the wake of the melting ice sheets, where the ground was continuously sodden with ice-cold meltwater.

Muskeg differs from "ordinary" forest in two very important respects. The first difference is this: in muskeg, the rate of accumulation of dead plants and plant parts exceeds their rate of removal by decay or fire. Bacterial and fungal decay are slow because of the low

Sphagnum

⌞ ⌟
1 mm

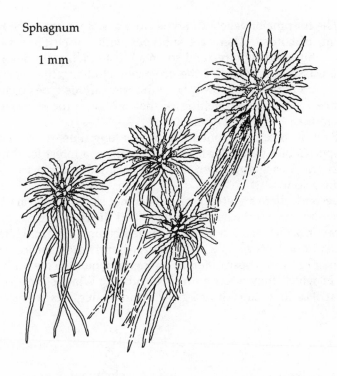

temperatures, and fires are rare because of the wet conditions. There-fore, and this is the most remarkable property of muskeg, it is an ecosystem that stores energy in the form of accumulated combustible (hence energy-containing) material. This material is peat. In most ecosystems, decay or fire removes the remains of dead plants as fast as they appear, but not in muskeg. According to Jon Terasmae,[15] an authority on muskeg ecology, as much as seventy percent of the solar energy absorbed and used for photosynthesis by the plants in mus-keg is conserved in peat.

The second remarkable property of muskeg is that, like a great sponge, it absorbs and accumulates water. Water enters a muskeg forest through precipitation (rain and snow) and through the ground (rivers, streams, and seeps). It leaves by evaporation (chiefly of vapor transpired by the plants) and by outflow through or over the ground. However, input and output are not in balance. Water accumulates and is held absorbed in the accumulating peat. One of the common-est plants of muskeg is sphagnum moss, otherwise known as peat moss; alive or dead, its water-holding capacity is renowned, and is what gives peat its great water-retaining power.

The four major vegetation types just described are by no means uniform, of course. There are subtypes within each type, subsub-types within each subtype, and so on almost indefinitely, depending on the classificatory zeal of the ecologists studying the vegetation. The question arises, however: How did the various types come into existence following the melting of the ice? The process must have been gradual, for several reasons.

First, it took time, sometimes a very long time, for the constituent species of a particular kind of vegetation, a forest for example, to immigrate, each at its own rate, as explained in the discussion of plant invasion.

Second, there are ecological delays that slow the arrival of plant newcomers, even if the seeds are there and the climate is hospitable. The way in which eastern hemlock tarried behind eastern white pine was mentioned previously. White spruce and black spruce are another pair of trees whose different ecological requirements control the order in which they arrive in new territory.[16] White spruce consistently arrives first, since it can grow on inorganic, newly uncovered

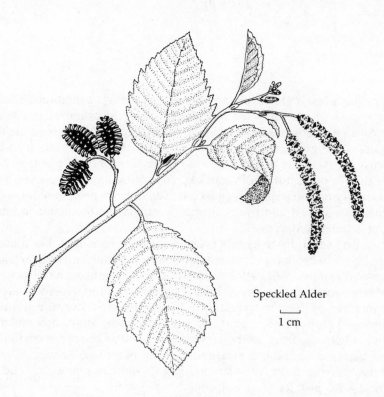

Speckled Alder

⊢—⊣
1 cm

soil; black spruce must lag behind, since it requires a spongy, organic soil where moss has accumulated.

What seems the appropriate order of arrival from the ecological point of view was sometimes reversed, presumably because of delays in migration. For instance, in "normal" ecological succession on newly available land, thickets of alder appear before forests of spruce. Alder is a nitrogen fixer and is intolerant of shade; therefore (usually) it comes first. It is then ousted by spruce, which can grow in the shade of the alders and which, by creating a deeply shaded spruce forest, prevents alders from continuing to flourish.

Events appear to have happened in this normal order in the revegetation of Labrador[17] but, surprisingly, not in New England[18] nor in northwestern Canada,[19] where alder arrived *after* spruce. Because alder cannot endure shade, the spruce that preceded it may have been present as open woodland rather than as dense forest. Another possibility is that alder invaded clearings in the spruce forests created by forest fires.

To return to the gradualness of vegetation change: slow environmental change and slow migration are only part of the explanation. There is also *ecological inertia,* a topic that merits a discussion of its own.

Ecological Inertia

Except for the earliest plant pioneers in newly deglaciated land, all later arrivals, if they were to succeed, had to take possession of ground already vegetated. They had to get started in gaps in the existing vegetation. But gaps would have been uncommon, and such gaps as there were would have received far more seeds from whatever plants happened to be abundant in the neighborhood than from the few, isolated specimens forming the vanguard of an invading species. This would have been true even if the climate were slowly becoming less suitable for the established vegetation and more suitable for the invaders. Hence the lag in vegetation change; once a plant species is abundant in a particular area, it can usually hang on there for a long time in spite of the climate's gradually becoming less suitable for it. Many plants have an astonishing ability to persist in unfavorable habitats.[20] The old adage "possession is nine points of the law" sums the matter up.

Such "logjams" often lasted until a fire destroyed a large tract of vegetation and left an opening. Then, provided the cleared area was large enough for the seeds of long-established plants to be as sparse as those of newcomers at its center, the newcomers would at last have a chance to gain a foothold. If they were better adapted to the current climatic conditions than were their predecessors, they would take possession of the land. They, in their turn, would persist until some time after the climate had become less than optimal for them.

Delayed responses caused damped responses. Because vegetation is slow to respond to climatic change, a short-lived change may not have time to produce a response. This may explain why, in some places, the vegetation changed steadily, without any reversals in the direction of the change, even though the dwindling of the ice sheets underwent several reversals. For example, the gradual disappearance of the Laurentide ice sheet was interrupted by a number of major readvances of the ice front in the southern part of the Great Lakes region, but there seem to have been no corresponding vegetation retreats.[22] The slowness of vegetation change is not the only possible explanation for this, however. The ice readvanced in great lobes (see chapter 1); one lobe, for instance, filled the whole basin of modern Lake Michigan. The lobes may have surged forward because of pressure from accumulating ice over the distant center of the sheet. There need not have been any climatic cooling at the ice front. Indeed, warm ground would have aided the lobes to slide forward by melting the undersurface of the ice and forming a lubricating layer of water. In any case, irrespective of whether the climate continued to warm up or cooled temporarily in front of the advancing ice lobes, the vegetation appears to have continued its development with no perceptible interruption.

The slow response of vegetation to climatic change has interesting implications. If climate changes continuously, as it appears to, then vegetation may never succeed in catching up with it. In the words of Margaret Davis,[23] plant (and also animal) communities are "in disequilibrium, continually adjusting to climate and continually lagging behind and failing to achieve equilibrium before the onset of a new climatic trend."

This opinion is not universal. The opposing point of view has been advanced by H. E. Wright, Jr., another leader in the field of Quaternary paleoecology. He assumes that vegetation and climate are at present in equilibrium[24] and describes ancient communities that had what appear (to us) to be mismatched mixtures of species as "disharmonious." The implication is that modern mixtures are harmonious. The argument in favor of this view[25] is that climate changes in stepwise fashion and the last step was taken a long time ago; therefore, because the climate has not changed appreciably for a long time, vegetation has by now had time to come into equilibrium with it.

There is a wealth of evidence,[26] however, showing that climatic change is never ending. Even if major climatic "steps" are comparatively quick, it is almost certain that the climate in the intervals between steps undergoes continual lesser changes. In the light of present knowledge, therefore, Davis's view, that disequilibrium in ecological communities is much commoner than equilibrium, is the more acceptable.

It should lead, in time, to a much needed change in popular thought. The notion espoused by so many nonprofessional ecologists—that the living world is "marvelously" and "delicately" attuned to its environment—is not so much a scientifically reasonable theory as a mystically satisfying dogma. Its abandonment might lead to a useful fresh start in environmental politics.

Photoperiodism

An apparent obstacle to long northward and southward migrations by plants is the phenomenon of *photoperiodism*. As is well known, many species of plants are genetically programmed to flower only when there are the appropriate number of daylight hours during a twenty-four-hour day. There are so-called long-day plants and short-day plants. Many arctic plants are long-day plants; they cannot flower until, as spring advances into summer, the length of the "day" (that is, the number of daylight hours) has reached the required minimum value for them, a value that differs from species to species. Even when the spring is abnormally warm, they cannot be "hurried," since their flowering is not controlled by temperature. Many temperate zone plants are short-day plants. Two that are well known to gardeners are chrysanthemums and Michaelmas daisies (asters); it is not the warmth of high summer that triggers their flowering but the shorter days of fall. They can be "deceived" into flowering ahead of time by deliberately reducing the number of hours in the twenty-four during which they are exposed to daylight.

This is the phenomenon of photoperiodism, and it has obvious implications for the northward and southward migration of plants in response to climatic warming and cooling. For example, when the climate becomes cooler, plants are forced (in anthropomorphic terms) to migrate southward to escape the cold of high latitudes. But this brings them to latitudes in which the longest days of summer are shorter than those in which they have been accustomed to flower. In theory, therefore, a climatic change could cause the climatic zone and the day-length zone to which a plant species is adapted to shift into nonoverlapping belts of latitude. Would the plant become extinct?

The answer appears to be no. Studies have been carried out on plant species found in both the arctic and the alpine tundra; their ranges extend from the high arctic, where the summer sun never sets, to midlatitudes in which the longest summer day is no longer than sixteen hours. Results show[27] that such species (if they are photoperiodic, which not all species are) have local genetic races, with

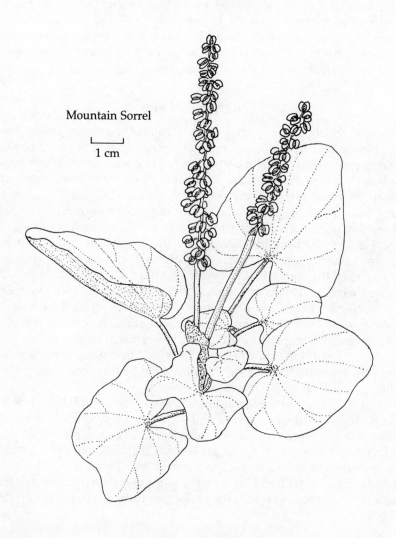

Mountain Sorrel

1 cm

each race adapted to the length of the summer days at the latitude where it grows. For example, mountain sorrel is one such plant. Presumably photoperiodic species with smaller ranges can also become adapted to latitude change if they have to. As such a species slowly migrates in order to stay in the climatic zone to which it is adapted, natural selection takes place; the survivors of every generation of seedlings tend to be those whose light requirements most closely match the light regime of the latitude belt where they grow.

PART TWO

THE TIME OF MAXIMUM ICE

5

Eighteen Thousand Years Ago: Life South of the Ice

We can now begin a chronological account of how life of all kinds invaded and developed as the great North American ice sheets dwindled until they almost disappeared. The time at which the total volume of ice began to shrink is hard to determine exactly but was probably close to 18k B.P. This date is often regarded as that of the final maximum of the Wisconsin glaciation. It should not be thought, however, that the margins of the ice sheets all retreated simultaneously in an orderly fashion; this was far from the case. The speed at which the ice disappeared varied from one place to another and from one time to another, and there were many temporary readvances.[1] Indeed, the ice continued to advance in some regions for thousands of years after it had begun to shrink elsewhere (see South Coast of Beringia, chapter 6). If a time-lapse moving picture could have been taken from a stationary satellite, the great ice sheets would have looked like active amoebas, with undulating outlines, wobbling unsteadily as they contracted to nothing.

The advance of the biosphere, as more and more ice-free land became available, must have been correspondingly irregular. It was not a continuous advance; there were innumerable delays, retreats, and readvances. Conditions for life close to the ice margins were sometimes harsh and sometimes benign but rarely constant for long. A general summary of biosphere history, because of its very generality, blurs the minor details and smooths the irregularities. Collecting and analyzing the data on which to base such a history, and arriving at general conclusions from a multitude of little bits of evidence, has kept paleoecologists occupied for decades, and the picture—a moving picture—becomes ever clearer. Much remains to be learned and disagreements over details abound, but an interim report on the highlights of what is now known is possible. It is best to begin the story with events in the region immediately south of the ice in what is now the northern United States.

Large Mammals and Their Environment South of the Ice Sheets

Conditions south of the ice sheets at 18k B.P. were briefly outlined in chapter 4. In the eastern half of the continent there was probably a mix of boreal spruce forest and tundra, but it seems certain that both communities were unlike their modern counterparts in many ways. In other words, they have no modern analogue. Then, as now, the climate became steadily drier from the Atlantic to the Rocky Mountains and, except in a belt immediately south of the ice, was too dry for forest to the west of modern Minnesota. Between the eastern forests and the mountains, in the midst of what are now are the grass-covered Great Plains, there was a vast area of sand dunes[2] of which the Nebraska Sandhills still remain. The sandstone bedrock of the area provided an abundance of sand, which was piled into dunes by the exceptionally strong periglacial winds (see chapter 1).

A detailed reconstruction of the vegetation in the periglacial zone south of the ice is not possible because data are inadequate.[3] Lakes with pollen-bearing sediments are common in regions that were once ice covered but much rarer in regions that were never under ice. Recent glaciation is the cause of the difference. Where ice sheets advanced over hard rock, as on the Canadian Shield, they

Mastodon

30 cm

gouged hollows that filled with water when the ice melted. Where ice sheets deposited sheets of till, as on the western plains, slow melting of blocks of stagnant ice left water-filled subsidence hollows or kettle holes (see Ice and Freshwater, chapter 1). The result in both cases is the so-called deranged drainage, which still exists: an array of small, local drainage basins that are not connected with one another because there has not been time since the ice melted for the streams and rivers to develop that will eventually link the lakes and drain them. These myriad separate lakes are the chief source of palynologists' sediment cores; where they are plentiful they provide data for a very thorough reconstruction of past vegetation.

South of the southern limit of glaciation nothing has happened to derange the drainage for several million years; except in desert country, all rivers lead to the ocean. Undrained hollows with their sediment-gathering lakes are scarce and far apart. As a result, past vegetation cannot be nearly so well reconstructed as it can on once glaciated land. Although vegetation maps have been compiled,[4] and certainly give the best possible approximation to past vegetation, they are based chiefly on interpolation inspired by educated guesswork.

There is no question, however, that there were forests, tundra, and grasslands of sufficient extent to nourish a tremendous assortment of large mammals.[5] The most famous of these were the elephantlike mastodons and mammoths. Mammoths (described in chapter 4) preferred comparatively open ground, especially tundra, and were commoner in the western half of the continent. Mastodons (sometimes written mastodonts) browsed on coniferous trees and seem to have been particularly numerous south of the Great Lakes and along the Atlantic coast; their chosen habitats were spruce swamps and, as second best, pine parkland.[6] In swampy environments they must have browsed chiefly on black spruce, but they are known to have eaten larch and hemlock as well, as fossil twigs of these species have been found inside the rib cages of fossil mastodons.[7]

Mastodons and mammoths were similar in many ways. Both were massive animals with huge, columnar legs and long, flexible trunks. They are therefore classed in the same zoological order, the Proboscidea. Early paleontologists emphasized their similarities and gave them confusingly similar generic names: *Mammut* for mastodons, and *Mammuthus* for mammoths. But because of pronounced skeletal differences between them, modern taxonomists have assigned them to different families: mastodons to the family Mammutidae and mammoths (along with modern elephants) to the family Elephantidae.

Mastodons had shorter, thicker limb bones than mammoths; they were hairy, but probably lacked the woolly undercoat of mammoths. They had straighter tusks, and the tops of their skulls were flat rather than crested. Mastodons were the more primitive animals, as shown by their less highly evolved teeth; mastodon teeth bore pairs of big, conical cusps, whereas mammoth teeth (like those of modern elephants) had a washboardlike chewing surface with numerous narrow ridges. Only one species of mastodon is recognized, the American mastodon. The mammoths were more variable and have been classified into four species,[8] only two of which were extant at the time the Wisconsin glaciation ended. The first, the woolly mammoth, roamed the tundra northwest of the ice in Beringia (see chapter 7), as well as south of it[9]; it was about the same size as a mastodon. The second was the considerably larger Jefferson's mammoth, otherwise known as the imperial mammoth, which lived south of the ice, chiefly in the center and west of the continent.

A multitude of other large mammals lived not far south of the ice sheets. As wide a variety of habitats must have existed then as now, each with its quota of species. Taking them all together, the diversity is impressive. Among the herbivores were several species that are still extant: wapiti (often called elk), white-tailed deer, mule

Shrub-ox

10 cm

Sabertooths

10 cm

deer, bison, and, in the western mountains, bighorn sheep. There
were also tundra muskox and caribou living far to the south of the
southern limits of their modern ranges, as noted in chapter 4. Moose
seem to have been absent; they are believed to have migrated south
from Beringia only after the ice sheets had begun to waste away.

Herbivores that are now extinct included Mexican horse, a small
horse that was probably common in the Great Plains, and western
camel, which is believed to have grazed in large herds. These two
species were abundant and, judging from their extant relatives, were
animals of grassland rather than of tundra. There was also woodland
muskox, which was taller and thinner than the extant tundra muskox
and had the bases of the horn sheaths fused to form a single dome of
horn; shrub-ox, with a skull apparently adapted to smash against that
of rivals in head-to-head combat, as happens with modern bighorn
sheep; and stag-moose, which resembled modern moose except for
its elaborately forked antlers.

These herbivores were prey to an equally diverse array of carnivores, only a few of which have persisted to the present. Of extant
species, there were timber wolves, though they were probably uncommon. Among the cats were cougars and bobcats. There were also
black bears, which are omnivorous rather than carnivorous, even
though for anatomical reasons they are classed as Carnivora. Grizzly
bears, like moose, probably reached midlatitude America only after
the ice sheets had begun to shrink.

The carnivores that are now extinct were far more spectacular. They included the dire wolf, which was similar to the modern timber wolf but much more powerful (see chapter 12); the famous saber-tooth, which was about the size of a modern African lion and had enormous saber teeth, apparently adapted for stabbing; the somewhat smaller scimitar cat, whose comparatively short scimitarlike canines had sharp, saw-toothed edges fore and aft; the American lion, and the American cheetah, which were not unlike their modern African counterparts. The giant short-faced bear may well have been the most awe-inspiring carnivore of them all; it was as tall as a moose, and compared with modern bears of the genus *Ursus*, rather short bodied, which suggests that it was a swift runner. Its short, broad muzzle must have enabled it to grip and hang on to its quarry (see chapter 12).

Human Life South of the Ice

The question of when human beings first settled temperate North America has been a subject of controversy for decades. That they had arrived and established themselves by about 11.5k B.P. is beyond doubt. This date marks the beginning of what has become known to archaeologists as Clovis culture. Humans were probably fairly numerous all across the continent, from Nova Scotia on the Atlantic coast to California on the Pacific, but in spite of the enormous size of the area, they appear to have shared a common culture. They all made fluted stone spearheads of the same distinctive design; the flutings make the base of a spearhead thin enough for it to be conveniently wedged into the split end of a wooden shaft. The oldest spearheads discovered are known as Clovis points; their makers and users are called Clovis people. To repeat: it is certain that humans have lived in temperate North America since 11.5k B.P. What is *not* known is whether there were people before them, sharing the continent with the multitude of other large mammals previously described.

There are plenty of signs of pre-Clovis human presence, but they are all equivocal. Possible indications have been found at twenty-six sites at least,[10] but in every case the evidence is dubious. Three separate flaws can invalidate archaeological evidence. First, there may be errors in the radiocarbon dating of bones or other organic remains (see chapter 2). Second, naturally broken rocks may be mistaken for human artifacts. Third, if a genuine stone artifact is mistakenly assumed to be contemporary with a sample of organic material found with it, the inferred date is, of course, wrong; the artifact may simply have been redeposited beside the organic material (a redeposited artifact is one that has moved, for example, by rolling downhill from the place where it was last used).

One or another of these problems clouds all pre-Clovis remains so far known and raises an interesting philosophical question: Is a large amount of dubious data more dubious or less than a small amount? Given a large quantity of data, surely some of it would be expected to be good.

If humans were indeed present before the Clovis period, two questions arise that are difficult to answer. Why were they so backward? And why were there so few of them? The first question arises because humans in the Old World have been expert makers of stone tools and weapons for at least 125,000 years,[11] whereas the things constructed by pre-Clovis people in the New World (if there were any) are so crude that there is some doubt that they are human artifacts. The second question is prompted by the fact that modern humans are so excessively prolific; why did the pre-Clovis population (if there was one) remain so small that no unquestionable signs of it remain?

The arguments against the existence of pre-Clovis people, taken all together, seem fairly strong, yet their existence is impossible to disprove and the possibility of a definitive discovery showing that they *did* exist will always be there, which makes it worthwhile to provide some details of a tantalizing discovery that has turned out to be as dubious as all the others.

In 1961 parts of a human skeleton said to be "at least 18,000 years old"[12] were found in alluvial sand in the valley of the Oldman River in Taber, Alberta. What may be human artifacts of the same age were found not far from Taber, in Medicine Hat, Alberta (see fig. 5.1), together with numerous bones from dire wolves, sabertooths, mammoths, camels, and asses.

The Alberta finds are interesting. The human remains, now known as the Taber child, were unearthed by a party of geologists led by A. McS. Stalker, the foremost pioneer in the unraveling and interpreting of the Quaternary geology of much of western Canada. They were the bones of a four-month-old baby just starting to grow his (her?) first molars; parts of the skull, a collarbone, part of a shoulder blade, a fragment of femur, two ribs, and two crushed vertebrae were recovered. The quantity of organic material was insufficient for the carbon-dating techniques used in the 1960s; however, the fossil was overlain by ten meters of alluvium topped by a sheet of glacial till, which suggests that it must have been in place since considerably before the last advance of an ice sheet over the site. From his knowledge of the history of ice movement in the region, Stalker put the

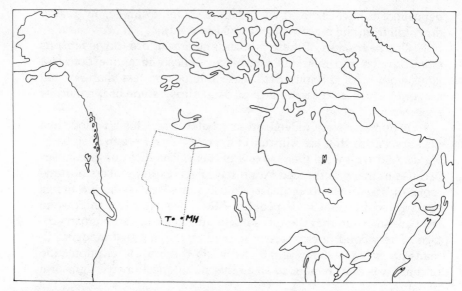

FIGURE 5.1: Site of Taber child finds. (T, Taber; MH, Medicine Hat.)
The dotted line shows the province of Alberta.

probable date of the fossils as 30k to 40k B.P., or rather less probably
at about 20k B.P. They could even be very much older, with a date of
60k B.P. The uncertainty arises because the southern margin of the
Laurentide ice sheet shifted back and forth repeatedly, over various
distances, throughout the Wisconsin glaciation. Each advance is
marked by a layer of till, and at a few sites it has been possible to
date the successive till layers by the radiocarbon method. At other
sites, especially those beyond the range of some of the ice advances,
it is difficult to judge which of the dated till layers each local layer
corresponds with.

More recent investigations of the Taber child's bones, using
modern radiocarbon methods, imply that the skeleton is only about
4,000 years old [13]; this assumes that the dated material was not contaminated with modern preservatives, which the investigators say
they took great pains to remove. Moreover, in the opinion of some
geologists, the bones could have been carried down and deposited
under the overlying alluvium by mudflows.[14] Other evidence, however, still points to the child's death as having occurred before the last
glacial advance, that is, before 18k B.P.

The date of Taber child is therefore still in doubt. And there are
other signs of early human presence, namely, the human artifacts (if
that is what they are) found at nearby Medicine Hat. They date from

between 17k and 22k B.P., so if the latest date given by Stalker for Taber child is the correct one, the child and these artifacts were more or less contemporary. Debate over the reality of the so-called artifacts demonstrates the contrasting approaches of archaeologists and geologists to flaked or fractured stones.[15] The stones in this case are pieces of chert, which were either flaked by human toolmakers or fractured by natural causes. The question is, which?

Like all scientists, archaeologists and geologists adhere to the principle of Occam's razor, according to which unnecessary assumptions must always be avoided. Therefore, given some pieces of fractured chert that might have been fractured by human hands, an archaeologist will not accept this possibility without supporting evidence; if the cherts could conceivably have been fractured naturally, then (to an archaeologist) they are inadequate evidence, by themselves, of human presence.

Given the same pieces of fractured chert, a geologist, equally mindful of Occam's razor, would approach the problem from the opposite direction. The question becomes: Of all possible causes for the fracturing, which is the least improbable? In the case of the Medicine Hat cherts, natural fracturing appears exceedingly unlikely; no geological explanation for it is tenable. Therefore, the only reasonable conclusion for a geologist is that the fracturing is human handiwork.

Even if we accept that human beings were present south of the ice sheets as early as 18k B.P., they were certainly not an important element in contemporary ecosystems. Pride of place goes to the large mammals described previously, of which the majority are now, unfortunately, extinct. The cause of their extinction is a topic of great controversy, to be discussed in chapter 12.

Plants South of the Ice Sheets

We know that many species of large mammals lived south of the ice 18,000 years ago and are now extinct. What about plants? Has there been a similar wave of extinctions among spectacular plants? The question seems not to have been asked, perhaps because it is impossible to answer. All mammals have fossilizable bones, but comparatively few plants have readily fossilizable hard tissues. Woody plants, ranging in size from trees to dwarf shrubs, leave wood fragments and twigs; among other hard plant parts are seeds, cones, some fruits, and leathery leaves. Wind-pollinated plants leave copious pollen. Even so, most herbaceous plants probably decay without trace; soft, succulent plant tissues seldom form fossils. Fewer plant species leave macrofossils than leave fossil pollen and only about five percent of the flora of a region is believed to be represented by the pollen in lake sediments (see chapter 3). Therefore, the forests, meadows, and plains where mammoths and mastodons grazed and browsed, and where sabertooths and dire wolves stalked their prey, may have contained some spectacular flowering plants that no longer exist. We shall never know.

Some plants appear to have become extinct over parts of their

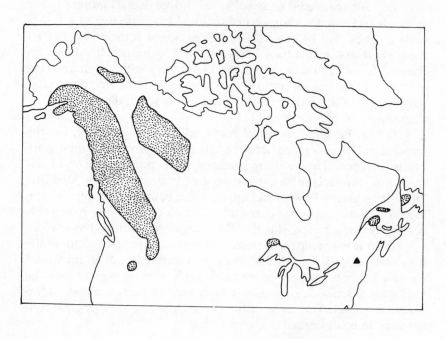

1 cm

FIGURE 5.2: The present ranges of two disjunct plant species. Left, yellow dryad. Right, the willow *Salix vestita*. The small black triangle shows the site in Vermont where numerous macrofossils of tundra plants, dating from about 11.5k B.P. have been collected.[17]

ranges. These are the disjunct species (discussed in chapter 3) that are found mainly but not exclusively in mountain habitats on the western and eastern sides of the continent, but not between. Many such species exist, and it is convenient to call them mountain disjuncts. The ranges of two of them, yellow dryad and the willow *Salix vestita*, are shown in figure 5.2. Both are species of open ground in cold climates and are found in mountain valleys at various altitudes.

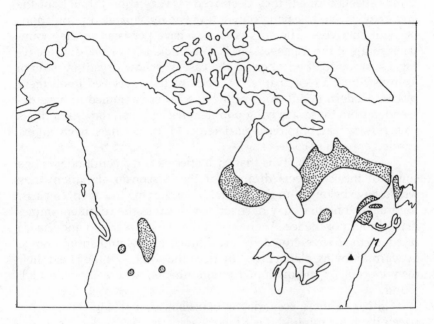

Presumably both species once had single, gap-free ranges and subsequently became extinct in the central parts of their ranges. Opinions differ[16] on when the extinctions happened.

One possibility is that these species had continuous ranges right across arctic Canada and south in the mountains on both sides of the continent before the Wisconsin glaciation, that is, during the preceding interglacial (the Sangamon Interglacial). Then, when the Wisconsin ice sheets formed, they destroyed all vegetation throughout the area they covered, leaving only a few widely separated populations of plant survivors. The survivors may have persisted in the mountains south of the ice sheets, in isolated nunataks or low-altitude refugia surrounded by ice, or (in the case of yellow dryad) in parts of ice-free Beringia (see chapter 1). Perhaps they survived in all these places. Modern populations of the species are assumed to have descended from these surviving populations; although the descendants have now spread onto once-glaciated land, their ranges have not expanded enough to coalesce.

Another possibility is that each species had a continuous range south of the ice sheets throughout the Wisconsin glaciation; they would have belonged to the vegetation occupying the open, treeless, tundralike zone that may have stretched across the continent immediately south of the ice. Then, when the climate warmed and the ice receded, they were crowded out of much of this midlatitude range by warmth-loving plants invading from the south; but they held their own in cool or mountainous environments, where they are still found.

Either of these scenarios is plausible, though the contrast between them is profound. One theory assumes that the localized "extinctions" producing the disjunctions happened more than 100,000 years ago and were caused by expanding ice sheets; the other, that they happened less than 10,000 years ago and were caused by competition with spreading warmth-loving plants. It is known that both yellow dryad and Salix vestita once grew at low elevations far south of their present range limits in what is now Vermont.[17] Fossil leaves of both species, dating from about 11.5k B.P., have been found there, at a valley site that was ice covered at the time of glacial maximum. It therefore seems likely that at glacial maximum these species lived even farther south and that they followed the receding ice front northward. Such a conclusion is consistent with either of the scenarios outlined.

There is no reason to believe that all the known mountain disjuncts (there are many) have the same history. The local extinctions that created gaps in their ranges may have happened at different

Lapland Diapensia

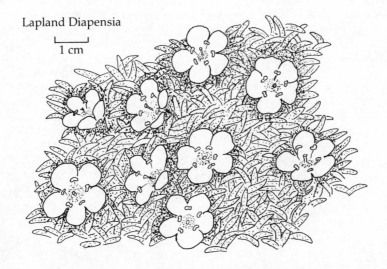

1 cm

times in different species. The longer two populations have been separated from each other, the more time they will have had to diverge genetically. Populations that have been disjunct since the beginning of the Wisconsin glaciation must differ from each other much more markedly than those that became disjunct only at the end of the glaciation. An example of a disjunct species whose eastern and western populations are treated as separate subspecies is Lapland Diapensia; the subspecies differ from each other in the shapes of their leaves. However, there is no way of deducing the duration of the disjunction from the magnitude of the morphological difference.

The various ways in which the mountain disjuncts could have developed their disjunct ranges have been debated for decades. No less interesting, though not so often discussed, are the ranges of plants that are *not* disjunct. Figure 5.3 shows two examples, mountain fireweed and mountain goldenrod. Both have what has been called a "rainbow" range[18]; that is, their ranges stretch right across North America at high latitudes and extend southward in the mountains in both the east and the west, giving each a range shaped more or less like an inverted (and tilted) V.

Such ranges inspire several questions: How did these species recover from the effects of the Wisconsin glaciation? From what sources and by what routes did they reach arctic and subarctic Canada once the ice sheets had melted? Were their ranges disjunct for a period, with subranges that spread and coalesced only recently? Did they survive south of the ice? Or in Beringia? Did they invade from Siberia, which was much less extensively glaciated than North Amer-

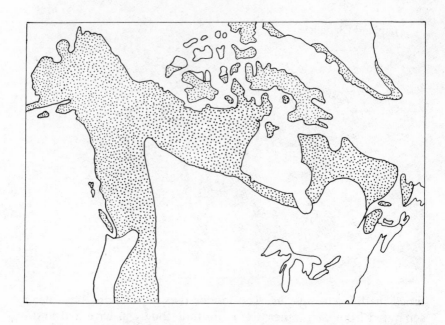

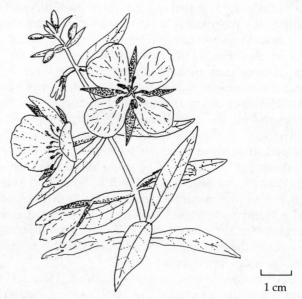

1 cm

FIGURE 5.3: The present ranges of two mountain and northern plant species that are not disjunct. Left, Mountain fireweed. Right, Mountain goldenrod.[19]

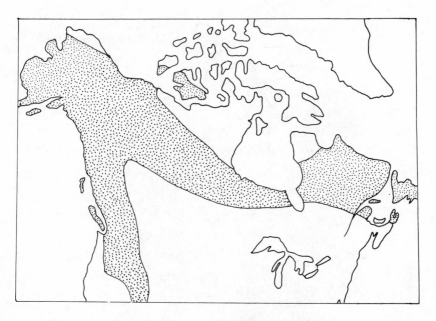

Golden Saxifrage and
Melissa Arctic Butterfly

⊢——⊣
1 cm

ica? None of these questions can be answered except, perhaps, the last. Mountain goldenrod is currently found only in North America, so it is most unlikely to have invaded from Siberia; the opposite is true of mountain fireweed.

Some butterfly species have rainbow ranges too; the development of their ranges presumably parallels that of the plants. An example is the Melissa arctic, known also as the White Mountain butterfly because within the United States it is found only in the White Mountains of New Hampshire.[20]

Many arctic and alpine plants with rainbow ranges in North America also grow in similar habitats in the Old World. For example, golden saxifrage grows in Greenland, Iceland, Great Britain, and Scandinavia; mountain sorrel (see chapter 4) grows in the high north latitude lands all around the pole. Both these species could, therefore, have invaded North America from overseas after the ice had melted. Other species with rainbow ranges are confined (or almost confined) to North America; for example, the range of white dryad (see chapter 3) is wholly North American except for a small outlier in extreme northeastern Siberia just across Bering Strait from Alaska; perhaps it survived in ice-free Beringia. However, fossils of all three of these species were found in the fossil deposits in Vermont mentioned previously, so they could equally well have spread into their present ranges from south of the ice.

All that can be said with certainty is that the geographical ranges of all species of plants are changing all the time. Conclusions about the history of these changes are speculative at best. One conclusion we can draw, however: the plant communities that gradually came into being after the Wisconsin glaciation, and that clothed what had been bare, sterile ground with a mantle of vegetation, continue to gain and lose species. The extent to which the glaciation has left us impoverished, in the sense that many of the species lost have not been regained, is impossible to judge.

6

The Coasts

The presence of great ice sheets was only one of the contrasts between North America at the time of the last glacial maximum and North America as it is today. Equally important from the ecological point of view was the existence of large tracts of dry land seaward of the modern coastline (fig. 6.1).

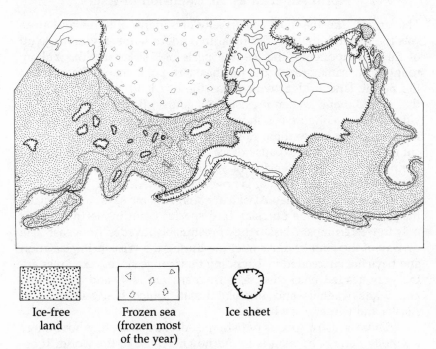

Ice-free land	Frozen sea (frozen most of the year)	Ice sheet

FIGURE 6.1: The pattern of land, sea, and ice at glacial maximum. The fine lines show the present-day coastline.

As noted in chapter 1, the Wisconsin ice sheets contained so much of the world's water that the volume of the world ocean was appreciably less than at present. Consequently, sea level was about 100 meters lower than it is now, and the area of the North American continent was considerably greater. Strips of what is now submerged continental shelf were then coastal plains.

It might be thought that if we had exact bathymetric maps (that is, relief maps of the ocean floor) and if we knew, precisely, the amount of the worldwide (eustatic) lowering of the sea level at the time of glacial maximum, we would be able to draw a fully accurate map of the coastline at that time. This is not possible, however. Allowance would also have to be made for isostatic sea level changes, that is, local changes resulting from warping of the earth's crust.

The map in figure 6.1 is therefore only approximate, but it does make clear one indubitable fact: both to east and west of the great North American ice sheets, there were extensive tracts of ice-free land that are now submerged beneath the sea.

North America as an Extension of Asia

The largest, and the most ecologically important, of these land areas was the region known as Beringia. In modern terms, it was made up of three components: the ice-free parts of Alaska and the Yukon; roughly the same area of the northeastern peninsula of Siberia; and the Bering land bridge, which was the dry land, now submerged, that united them.[1] When it existed, of course, Beringia was an undivided whole, as figure 6.1 shows. There was no North America and no Asia as we know them now, only a single, enormous continent cut in two by a great expanse of ice. Immediately west of the ice was Beringia; it was an extension of what is now Asia. South of the ice was the ice-free part of what is now North America.

Parts of present-day Alaska are still, in some says, more Asiatic than American. Some Eurasian bird species occur in North America only in westernmost Alaska; they presumably invaded from east Asia by way of the land bridge sometime during the Wisconsin glaciation and have not succeeded in extending their range since.[2] Examples are the gray-headed chickadee, a permanent resident, and the yellow and white wagtails and the bluethroat, which breed in western Alaska and winter elsewhere.

There is also a species of fish, the Alaska blackfish, whose range is wholly Beringian,[3] as it is found nowhere else in the world. It occurs in the Chukotka Peninsula, in the Amguema River, on Saint

Gray-headed Chickadee

1 cm

Yellow Wagtail

1 cm

Alaska Blackfish

1 cm

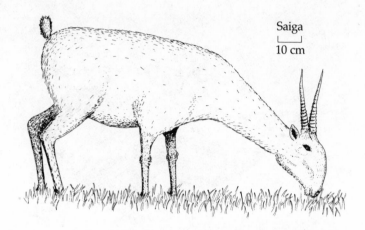

Saiga
⊢——⊣
10 cm

Lawrence, Saint Matthew, and Nunivak Islands in the Bering Strait, and in western Alaska (fig. 6.2). At the height of the Wisconsin glaciation, the three islands were merely hills in the low plains of the Bering land bridge.

Not all the Asiatic animals that invaded eastern Beringia (or northwestern North America as it is now) are still to be found there. For some species, fossils are all that remain. An example is the saiga, which is now confined to the steppes of central Asia. Saiga fossils have been recovered near Fairbanks, Alaska,[4] and in the Baillie Islands[5] (fig. 6.2).

Saiga was not the only large mammal to cross the land bridge from Asia. As we shall see in chapter 7, several other species came as well. But saiga is the only one that failed to extend its range southward. At the present time it is thought of as an exclusively Asiatic animal; its presence in "North America" or, strictly speaking, in Beringia, during the last glaciation emphasizes that Beringia was ecologically a part of Asia.

The ecology of Beringia as a whole and its links with land south of the ice sheets are the topics of chapter 7. Subsequent sections of this chapter deal with coastal paleoecology, east and west of modern North America.

The South Coast of Beringia

Any attempt to visualize conditions on the Pacific coast of Beringia at the time of glacial maximum must begin with thoughts about what the climate would have been like. The whole coast was strongly affected by the huge ice sheet to the east and by the absence of a sea link between the Pacific and Arctic oceans.

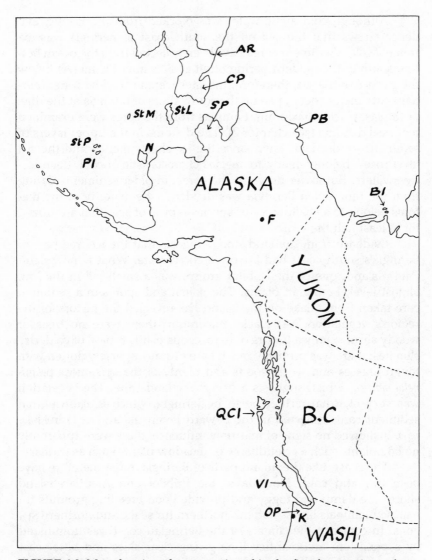

FIGURE 6.2: Map showing places mentioned in the first four sections of chapter 6. Abbreviations: AR, Amguema River; BI, Baillie Islands; CP, Chukotka Peninsula; F, Fairbanks; K, Kalaloch; N, Nunivak Island; OP, Olympic Peninsula; QCI, Queen Charlotte Islands; SP, Seward Peninsula; St L, Saint Lawrence Island; St M, Saint Matthew Island; St P, Saint Paul Island; VI, Vancouver Island; PI, Pribilof Islands.

Although cold arctic water was blocked from the Pacific, evidence shows that the sea off the south coast of Beringia was extremely cold.[6] Sediment cores have been collected from the ocean bottom south of the modern Bering Strait at sites now 150 meters below the surface of the sea; therefore, the sites were in shallow water (perhaps fifty meters deep) just off the south shore of Beringia at the time of the last glacial maximum. Samples from the cores were examined for fossil diatoms (see chapter 2). The diatoms in the upper layers of sediment—those laid down since the ice sheets melted and the sea level rose—belong chiefly to species characteristic of cool, deep ice-free waters. But lower down in the cores, in older sediments dating from the time when Beringia was dry land, the majority of the diatoms belong to a shallow water species typical of seas that are frozen for at least half the year.

Evidence from the land, too, suggests that the ice-free part of Beringia's south coast had a truly arctic climate. What is now Saint Paul's Island, one of the Pribilof group, was a small hill in the low, almost level, Beringian plains. The pollen and spores in a sediment core taken from a lake on the island has revealed the history of the region's vegetation.[7] At glacial maximum, there were no trees or woody shrubs, only a tundra of herbaceous plants typical of cold, dry climates; there was probably much bare ground, sparsely dotted with tough grasses and wormwoods and plenty of the spikemoss *Selaginella sibirica*, which suggests a dry, rocky landscape. The vegetation was very like that farther north in Beringia, which is known from sediment cores collected in the Seward Peninsula and at Point Barrow. It showed no signs of maritime influence: there were apparently no bog plants such as cloudberry or meadow plants such as fritillary.

The coast, like the inland parts of Beringia, must therefore have been dry and cold. Nowadays, the Pribilof and Aleutian Island shores are damp and foggy and provide good breeding grounds for marine mammals like sea otters, northern fur seals, and northern sea lions. In contrast, the climate of the Beringian coast was dominated by a persistent arctic anticyclone giving clear skies and bitter east winds off the Cordilleran ice sheet, conditions inhospitable to marine mammals that bear their young on shore or on ice floes.

The Western Edge of the Ice

The ice-free part of the Beringian south coast was separated from the west coast of unglaciated North America by the long stretch of shore where the Cordilleran ice sheet reached the sea. Much of this coast must have consisted of high ice cliffs, but not all of it. Numerous ice-free refugia are believed[8] to have been scattered along the shores and on the offshore islands of what are now Alaska and British Columbia. These refugia would have been important as stepping stones in the migration of living things (including humans) at a time when all other routes between Beringia and ice-free North America were blocked by ice. Before considering the ecology of this coast, however, a few words are necessary on its complicated geological history.

The Cordilleran ice sheet was still spreading westward for some 3,000 years after the Laurentide ice sheet had begun to shrink. Indeed, although the North American ice sheets treated as a whole reached maximum size about 18k B.P., the Cordilleran sheet lagged, reaching its maximum about 15k B.P. About this time there must have been a long period during which the ice front made numerous local advances and retreats as ice lobes surged seaward and then wasted away by calving off bergs. Probably many of the lobes flowed as valley glaciers between the high coastal mountains; separating these glaciers were bare, ice-free ridges.

While all this was going on, sea level was varying in a complicated manner.[10] How it varied at any point was the result of three separate causes. First, there was a gradual eustatic rise as water flowed from the shrinking ice sheets into the world ocean. Second, there were isostatic changes caused by the sagging of the earth's crust under the overlying load of ice. Third, there was another kind of isostatic variation due to *diastrophism* (the deformation of the crust through folding, faulting, buckling, and the like) caused by events in the earth's interior, especially by drifting of the crustal plates. This third cause of sea level change was probably less important than the other two but appreciable nevertheless. It was, and is, far more important on the Pacific than on the Atlantic coast of northern North America because crustal plates are colliding just off the Pacific coast.

The Cordilleran ice sheet thinned rapidly toward its seaward margin. At its westernmost edge, in the region of the Queen Charlotte Islands and the outer coast of Vancouver Island, it was quite thin and therefore light in weight. Consequently, at glacial maximum, the eustatic fall in sea level was greater than the isostatic rise, and land now submerged was above the sea, forming temporary sea level refugia. There were also mountain refugia (nunataks) where mountain summits rose above the comparatively thin ice sheet. A short distance to the east, where the ice sheet was thicker and heavier, the isostatic rise in sea level was greater than the eustatic fall, so that sea level along what is now the mainland coast was higher than it is now.

There must have been a great variety of landscapes. For any given point in low-lying parts of the coastal region, at any given time,

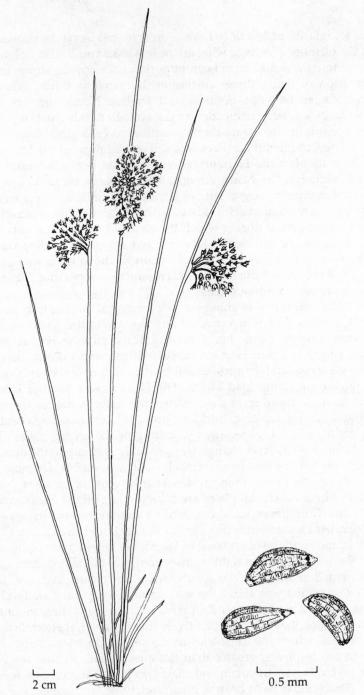

2 cm

0.5 mm

A rush (Juncus) with seeds

we can envisage four possible conditions: landfast ice, a floating ice shelf, ice-free land, or open sea. Perhaps many places experienced all these states several times over. In the mountains there was variety, too. Although most of the mainland was under ice, there were protruding summits (nunataks both temporary and permanent), bare ridges between glaciers that flowed into the sea, and bare cliffs where the ground was too steep for ice to cling. How large and how numerous these various refugia were is, of course, unknown.

A lowland refugium has been found on the east side of the Queen Charlotte Islands.[11] The evidence consists of fossil seeds found in layers of buried sediment about three meters from the top of a sea cliff. There are plentiful seeds of rushes, dock, pondweed, and a member of the chickweed family, as well as stonewort nucules (see Microfossils, chapter 2). They date from 15k or 16k B.P., when the Cordilleran ice sheet had reached its westernmost limits, and plant fossils in the overlying sediments prove that the area has been ice free ever since. It has not been possible to date the glacial till underlying the oldest seed-bearing layer, but the site has been a refugium for at least 16,000 years and probably much longer, as indicated by the varied and abundant vegetation that was obviously present and would have taken considerable time to become established.

Far to the south, near Kalaloch, Washington, is another sea cliff with pollen-bearing sediments.[12] The site is just beyond the farthest limit of the Cordilleran ice sheet and was probably not cut off from the unglaciated lands to the south even though it was hemmed in between the sea and the ice cap covering the Olympic Mountains. The pollen record spans an interval of more than 60,000 years, from about 70k to 17k B.P. It shows that the vegetation changed repeatedly, switching back and forth from forest to tundra and back to forest over and over again, no doubt in response to climatic fluctuations. A forest of western hemlock and spruce (probably Sitka spruce) grew in the region in mild periods and was replaced by a tundra of grasses, sedges, and various herbs in the cold periods.

The Kalaloch record emphasizes, with concrete evidence, a fact that is obviously true: conditions in glacial refugia varied continuously, and the plants and animals that persisted in them had to endure a wide range of climates. It does not follow, however, that the ecosystems in the refugia responded in the same way as they would today. This is especially true of isolated refugia at a great distance from unglaciated parts of the continent, which Kalaloch was not. The range of conditions—the niche—that an organism occupies is nearly always less than the range it could occupy if there were no competing organisms battling for possession of the same territory. Therefore,

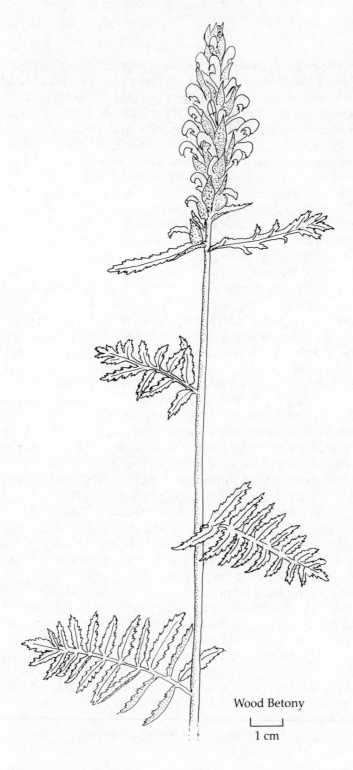

Wood Betony

1 cm

even if the climate of a high-altitude, tundra-covered nunatak became warm enough for tree growth, trees could not have grown if there were no tree seeds. In the absence of trees to crowd them out, the established tundra plants would have continued to flourish. The phenomenon of ecological inertia (see chapter 4) must be more pronounced in isolated "islands" of vegetation than in continent-wide tracts.

Most of the evidence for west coast refugia comes from peculiarities in the present-day geographical ranges of various organisms. Because the ice over the Queen Charlottes and western Vancouver Island was comparatively thin, many mountain summits were left uncovered[13] and could have provided safe havens for alpine and subalpine plants. Several species of plants are represented in these small, high-altitude refugia by subspecies that grow nowhere else and that presumably evolved their distinctive peculiarities during millennia of isolation.[14] A subspecies of clubmoss moss heather is thought to be found only on mountains in the Queen Charlottes and Vancouver Island. Subspecies of rosy paintbrush, western spring beauty, and wood betony are believed to occur only on mountains in Vancouver Island and the Olympic Peninsula.

The plants just mentioned are alpines, and their refugia were nunataks. There is a contrast worth noting between high altitude nunataks and shoreline refugia: whereas mountain summits stay fixed while ice sheets flow around them, shoreline refugia, which are ice free merely because the spreading ice happened not to reach them (see chapter 1) may shift their location rapidly as one ice lobe retreats and another advances. Indeed a refugium consisting of a gap between sea and ice must shift (or widen, or disappear) as sea level falls and rises. If, as seems likely, there were many of these ephemeral shoreline refugia along the north Pacific coast, they would have provided stepping stones along which shore plants could migrate northward and southward.

The present-day ranges of many west coast plants suggest that they spread along the shore in this manner. The modern flora of the Queen Charlotte Islands provides examples.[15] Some shore plants now found there must have immigrated from Beringia, for example, beach lovage, beach senecio, and oysterleaf. Others, such as yellow sand verbena, large-headed sedge, gray beach pea, and coast strawberry, immigrated from the south (fig. 6.3). But it should be recalled (see discussions of geographical range maps in chapter 3) that even when modern geographical ranges give fairly convincing evidence as to the route a migrating species followed, they tell nothing about when the migrations took place. The arrivals of these beach plants in

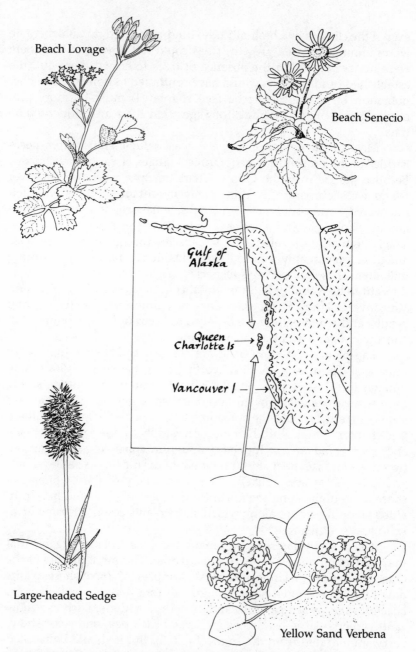

FIGURE 6.3: Some Queen Charlotte Island beach plants showing the directions from which they probably immigrated.[15] The shaded area on the map shows the maximum extent of Wisconsin ice.

River Otter

|___|
5 cm

the Queen Charlottes could have happened at any time; probably they are still migrating.

The shoreline refugia also provided stepping stones and at least temporary havens for mammals. Some mainland mammal species have still not reached the offshore islands, and perhaps they never will, in the present interglacial at any rate. Thus grizzly bear, fisher, and snowshoe hare occur on the mainland coast but not on Vancouver Island nor on the Queen Charlottes. Mink and mule deer are absent from the Queen Charlottes (except for introduced deer), though they are present on Vancouver Island. Among mammals now found on the islands, it is not known which are immigrants and which, if any, survived the Wisconsin glaciation in refugia. The most likely glacial survivors would have been animals such as river otters, martens, ermines, and black bears, which would have been able to travel over the ice from one refugium to another. In hospitable refugia, they could have lived on the beach, feeding on fish, crabs, and molluscs.[16]

Assuming then, as seems reasonable, that there was a chain of glacial refugia along the coast and that they were ecologically productive and stocked with a wide variety of plants and animals, it also seems likely that the refugia functioned as stepping stones for human migrants.[17]

Invaders from Asia are known to have traveled from Beringia to southern North America as the Wisconsin glaciation came to an end, but there has been much debate about the route they took. Many archaeologists assumed that they could only have come along the so-called ice-free corridor between the Cordilleran and Laurentide ice sheets, which was open for most, though probably not all, of Wisconsin time. Travel by this route would have been difficult and dangerous, however (see The Ice-free Corridor, chapter 7), and the coastal stepping stone route would have been much easier and more attractive.

The climate of the coastal route was probably fairly mild, at least along the more southerly stretches, because the sea was warmed by the Japan current, undiluted by arctic cold currents, which were blocked by the presence of Beringia. And it would have been foggy, because cold air off the ice sheet flowed over the warm sea. It is not known where along the shoreline the clear arctic cold of south coast Beringia gave way to the foggy warmth of what is now the British Columbia coast. South of the changeover, however, the refugia were probably covered with luxuriant vegetation and well supplied with meat: seals and sea lions, fish and shellfish, and migratory birds that would have used the chain of refugia as stopping points along a flyway. Caribou, a subspecies of which occurred in the Queen Charlotte Islands until becoming extinct in 1910, would also have been found there. The human invaders from Beringia probably found it easy to travel from one refugium to another, by boat in the summer and over the sea ice in winter. A coastal population of Paleo-Indians with a maritime culture may have become established independently of, and earlier than, the inland Paleo-Indian population of the Great Plains, which subsisted on herds of grazing animals, especially bison.

The East Coast Plains and Islands

Now consider the east coast. At glacial maximum, the sea surface was about 120 meters below its present level. Even so, North America and Greenland were not linked by a land bridge because the channel separating them is too deep. The two regions may have been joined by a narrow ice bridge between Ellesmere Island, the northernmost of the Canadian Arctic islands, and northern Greenland (see figs. 1.1 to 1.5). But this link, if it existed, had no biological significance.

The most important tracts of "extra" land, that is, land that was dry at glacial maximum but is now submerged beneath the sea, were at fairly low latitudes, from Newfoundland southward (fig. 6.4). They are now part of the wide continental shelf off eastern North America, but they were then part of the coastal plain.

South of Cape Cod the exposed coastal plain formed an unin-

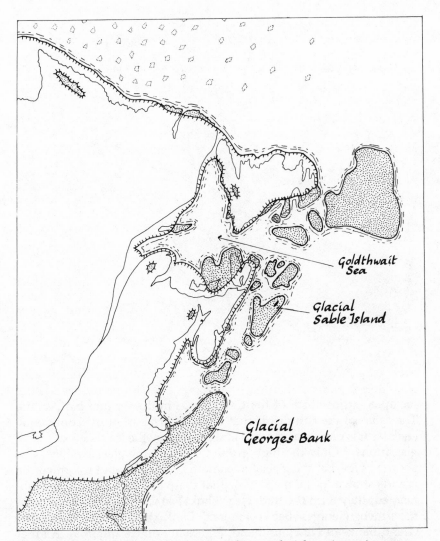

FIGURE 6.4: The pattern of land, sea, and ice at glacial maximum in eastern North America. This is an enlargement of the eastern part of the map in fig. 6.1. The modern Sable Island is the tiny black streak in Glacial Sable Island; it is all that now remains above water of the whole chain of large, low islands that lay offshore.

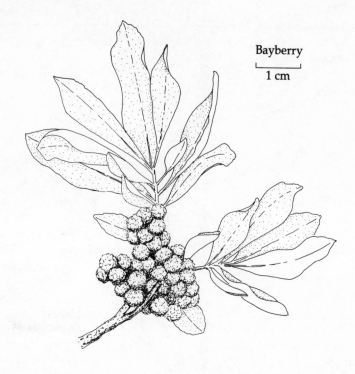

Bayberry
├──────┤
1 cm

terrupted strip; north of the Cape it was in the form of big islands. The cause of the contrast is that the northern part of the continental shelf (as it is now) was gouged and scoured by the ice sheets of earlier glaciations.[18] Only the high ground between the glacial valleys was above sea level at the glacial maximum of 18k B.P., and it formed the islands shown in the map. Now that these one-time islands are submerged, they form the undersea "banks" so well known to east coast fishermen: Georges Bank, Emerald Bank, Banquereau Bank, the Grand Banks of Newfoundland, and many others (fig. 6.5). All that now remains of them above sea level is Sable Island, which is no more than an arc of sand dunes forty-five kilometers long and a mere two kilometers wide at its widest point.

Fishermen trawling on Georges Bank for scallops, clams, and bottom fishes have contributed greatly to our knowledge of the region's paleoecology. They have recovered numbers of mastodon and mammoth teeth in their trawls.[19] This proves, first, that the animals were there when the bank was dry land; second, that the climate must have been suitable for them; and third, that the supply of plant food was adequate.

Mastodon teeth have been found in much greater number than mammoth teeth. This may imply that mastodons were more numerous than mammoths, or it may merely result from the greater strength of mastodon teeth, which are less likely to disintegrate than those of mammoths. In any case, the presence of mastodons almost certainly implies a supply of coniferous trees for them to browse on (see chapter 5). It is harder to deduce conditions from the mammoth finds because it is uncertain which species of mammoth the teeth come from; they do not completely match any known collections from other regions. They are comparatively small and may represent a dwarf variety of the woolly mammoth (see The Starting Conditions, chapter 4). If so, it seems reasonable to conclude that the vegetation consisted of a mixture of tundra (for the mammoths) and coniferous parkland and black spruce swamps (for the mastodons).

Northeastward along the chain of offshore islands, the trees may gradually have thinned out until, by the time Glacial Sable Island was reached, the vegetation consisted entirely of low shrubs and herbs.[20] Buried deposits of soil and peat have been sampled on modern Sable Island, and no sign of trees has been discovered—no stumps, no logs, and no wood fragments. The samples contain small amounts of pollen from pines, spruces, firs, birches, and alders but not more than would be expected if it were blown in from trees on the mainland.

At glacial maximum, Glacial Sable Island was huge, perhaps 250 kilometers long by 80 kilometers wide, and it seems to have been covered by much the same mix of vegetation as is now found on the tiny surviving remnant: beach grass and beach pea on sand hills, bayberry and withe-rod in moist hollows, and cranberry and crow-

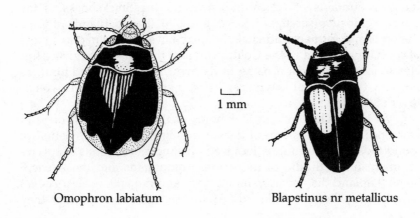

1 mm

Omophron labiatum   Blapstinus nr metallicus

berry where the soil was acid. Trees cannot grow on modern Sable Island, because wind-driven sand and salt spray quickly kill them. It is uncertain whether these same conditions or a colder climate prevailed when the island was large, making tree growth impossible.

Presumably the plant and animal inhabitants of modern Sable Island represent only a small fraction of the species that occupied the glacial island; as the sea rose and dry ground dwindled, many lineages must have gone extinct. Among probable survivors are several species of beetles that are nearly or completely incapable of flight.[21] For example, the two shown in the drawing (*Omophron labiatum* and *Blapstinus* nr *metallicus*), both weak fliers, are found on Sable Island even though the northernmost limit of their mainland range is Cape Cod (fig. 6.5). It seems likely that in the past their range stretched northward along the whole island chain; members of the species now alive on Sable Island are almost certainly marooned descendants of ancient, wide-ranging populations, for these beetles are probably incapable of crossing the broad stretch of ocean that now lies between island and mainland.

The East Coast Refugia

A variety of glacial refugia existed on and near the east coast, the most important being the islands and peninsulas, now submerged, described previously. The inhabitants of these coastal plain lands (as they will subsequently be called) were probably the ancestors of a large proportion of the plants and animals now living in New England and the Atlantic provinces of Canada: Nova Scotia, New Brunswick, Prince Edward Island, and Newfoundland. There were probably also small coastal refugia (see chapter 1) and nunataks in the mountains, for example, in the Torngats of Labrador, the Long Range Mountains of Newfoundland, and the Shickshocks of the Gaspé Peninsula (see figs. 1.8, 6.4, and 6.5). The highlands of Cape Breton Island formed a nunatak that may have been joined to a piece of coastal plain land in the Goldthwait Sea; the reason for giving this almost landlocked sea a name of its own, rather than calling it, as now, the Gulf of Saint Lawrence, is that there was no Saint Lawrence River at the time.

At many places along the shores of the modern Gulf of Saint Lawrence, and on the tops of the surrounding mountains, plants are found today that form rare, localized, disjunct populations of species common in the tundra of the western mountains and the Arctic.[22] Some (Lapland diapensia, yellow dryad, and the willow *Salix vestita*) have already been mentioned. At the last glacial maximum they would have had much more extensive eastern ranges than they have now if, as seems likely, there were big patches of tundra on the coastal plain lands. There is no need to assume that these currently rare and local plant populations survived the last glaciation only in

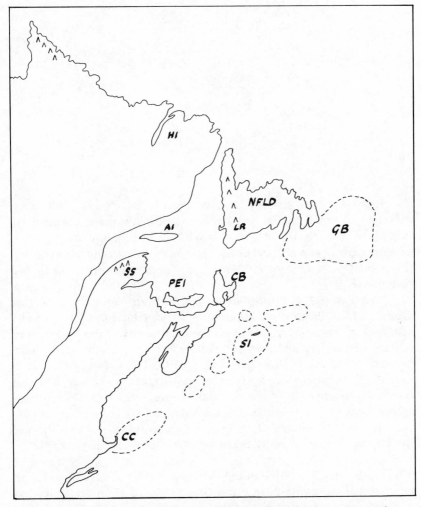

FIGURE 6.5: Map showing places mentioned in the last two sections of chapter 6. Submarine banks are shown with broken outlines. Abbreviations: NFLD, Newfoundland; SS, Shickshock Mountains; PEI, Prince Edward Island; CB, Cape Breton Island; SI, Sable Island; CC, Cape Cod; GB, Grand Banks of Newfoundland; HI, Hamilton Inlet; AI, Anticosti Island.

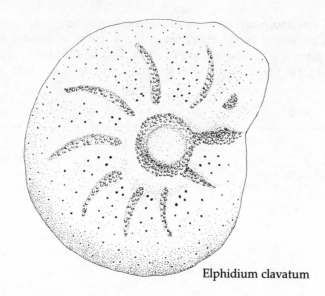

Elphidium clavatum

the sites where we find them now, even if these sites really were nunataks or small coastal refugia. Quite possibly these so-called refugia were *not* ice free at glacial maximum but merely lost their ice cover relatively early, thus becoming available for colonization before the coastal plain lands and their vegetation were submerged by the rising sea.

What needs to be explained (see Refuges from the Drought, chapter 13) is not how and where they survived the harsh conditions of the glaciation, but how and where they survived the benign conditions of the hypsithermal interval, when they could easily have been crowded out by the spreading forests. The answer probably is that they are plants of exposed, open habitats—sea cliffs, shifting river gravels, marshy ground, talus slopes, and mountain tops—where trees have no competitive advantage.

There is considerable disagreement about the exact position of the Laurentide ice sheet's outermost margin, as noted in chapter 1. The size, and even the existence of ice-free stretches along the coast of Labrador, has been one of the debatable points. There is reason to believe that ice-free land around the mouth of Hamilton Inlet has existed ever since 21k B.P. and that it was densely covered with tundra vegetation of willows, sedges, and peat moss.[23] Pollen and spores found in cores of marine sediment collected off the inlet provide the evidence. The region was not necessarily a refugium all through the Wisconsin glaciation. It could be a place where the ice sheet began its retreat unusually early; if so, it would be interesting to know the source of the spores and seeds that colonized it when it first became free of ice.

Foram shells (see chapter 2) were also abundant in the cores. The environmental requirements of the dominant species, *Elphidium clavatum*, are well known. It is a bottom dweller that lives close to the shore in waters that are ice free in summer. Pollen and forams together therefore imply that the climate was not excessively harsh.

Now consider the mammals that lived in these refugia. Not much is known about the subject apart from the fact that there were mastodons and mammoths on at least the more southerly of the coastal plain lands. Presumably the mammals now found in the region lived in or passed through the various refugia, and they must have been the ancestors of the current mammal inhabitants of what are now islands around the Gulf of Saint Lawrence: Newfoundland, Cape Breton Island, Prince Edward Island, and Anticosti Island. Present-day species lists for these islands inspire guesswork about the migrations that *could* have happened. Biological considerations

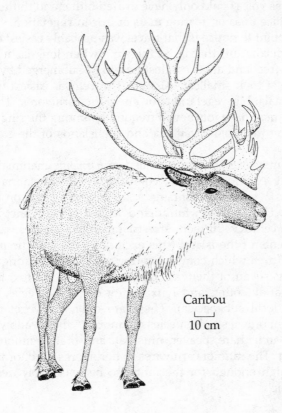

Caribou

$\llcorner\quad\lrcorner$
10 cm

tell us in general terms what was possible but not, of course, what actually took place.

Let us look at the possibilities. Whether or not a particular species of animal can survive for generation after generation in a refugium depends on the size of the refugium. Animals such as insects and other invertebrates, and small mammals such as mice and shrews, need only a few hectares of well-vegetated land to sustain themselves. But larger mammals must have larger territories. Large vegetarians such as caribou may need several hectares per head to meet their requirements for plant food; carnivores such as wolves, lynxes, and wolverines need extensive hunting territories. This is not to say that large mammals cannot make use of small areas; they certainly can, provided they can enter and leave them easily. And whether they can do this depends on the state of land and sea. Animals able to travel over landfast ice in summer and frozen sea in winter (for example, wolves) must have been able to travel at will around and across the Goldthwait Sea, and strong swimmers (for example, caribou) would have crossed easily from island to island. Small animals such as voles could only have drifted hither and thither unintentionally on ice floes or floating mats of frozen vegetation.

The point to notice is that travel was probably easiest at the time of glacial maximum, that is, at the time when long-lasting destinations—ice-free land areas *not* destined to be submerged by the rising sea—were at their smallest. Also, that travel was easiest for large or active animals: big vegetarians or energetic carnivores. These kinds of animals may have journeyed frequently among the ring of refugia surrounding the Goldthwait sea and the islands of the coastal plain lands.

This may explain the so-called *disharmonic* mammal fauna of present-day Newfoundland[24]; compared with the fauna of North America as a whole, it contains a surprisingly large proportion of carnivores. Until the Newfoundland wolf became extinct in the late nineteenth century, thirteen terrestrial mammal species were permanent residents of the island (this excludes bats, and the polar bears and arctic foxes which come as spring visitors on floating pack ice). As many as seven of them are carnivores: wolf, red fox, black bear, ermine, marten, otter, and lynx (many of these species are represented by local subspecies). There are only five vegetarians, and among them only a single species of "mouse," the meadow vole. The others are arctic hare, beaver, muskrat, and that strenuous traveler, the caribou. The ratio of carnivores to herbivores is abnormally high, certainly high enough for the fauna to be reasonably described as disharmonic.

7

Beringia and the Ice-Free Corridor

We now return to the west and to two great debates on how living things fared when the ice sheets were at their largest.

The first debate concerns Beringia, which persisted as a peninsula of Asia until about 15.5k B.P. when the rising sea first breached it to create Bering Strait.[1] The fossil record shows that many species of large mammals lived there, for example,[2] among vegetarians, woolly mammoth, mastodon, woodland muskox, western camel, Lambe's horse, and long-horned or steppe bison, all now extinct. Other, still extant, herbivores were also residents: caribou, tundra muskox, wapiti or elk, and Dall's sheep. Preying on them were short-faced bear, American lion, and the sabertooth cat, all now extinct, and the extant grizzly bear, wolf, and wolverine. And there were the omnivorous humans. Many of these animals lived south of the ice sheets, too (see chapter 5). The problem is: How did such large numbers of animals, many of them massive, sustain themselves in the spartan conditions of arctic and subarctic Beringia? How did they get enough to eat? Was Beringia truly populated with herds of big game?

The second debate concerns the so-called ice-free corridor that may have separated the Laurentian and Cordilleran ice sheets. It would have paralleled the Rocky Mountains, just to the east of them. The basic question is: Was there, or was there not, a corridor between the two ice sheets throughout the whole of the Wisconsin glaciation? A corridor certainly formed there when the two ice sheets began to melt, and conditions along it must have affected the migrations of plants and animals in both directions.

The following sections focus on these two debates and Beringia's human inhabitants.

Beringia and Its Big Game

Beringia was home to an impressive collection of large mammals. The preservation of their skeletons as good fossils, and often the finding of these fossils, result from two fortunate accidents. First, because Beringia was not overridden by ice, the fossils were less likely to be crushed. Second, Beringia includes the famous Klondike gold deposits in the bed of the Klondike River and its tributary creeks; placer miners thus unearthed many fossils that would not otherwise have been found. The finds are not only of bones, horns, and teeth: frag-

Tundra Muskox

└────┘
10 cm

Solifluction lobes

ments of the "freeze-dried" flesh of mammoth and horse have also been discovered.

The fossils are embedded in frozen *muck*, the technical term for the muddy sediment overlying the gold-bearing valley gravels. This muck reached its present position on top of the gravels by flowing down the surrounding hills. When the uppermost layer of soil on the slopes melts in summer, it forms a liquid mud saturated with water that cannot drain away because the underlying soil is frozen. The layer of mud therefore oozes down the frozen surface beneath, often forming lobes as it does so. The process is known as *solifluction*. The mudflow carries any contained fossils with it. In this way, the fossils are brought together in the valleys, where they accumulate.

The fossil finds suggest a wealth of animal life in ice age Beringia. Mammoth, horse, and bison were everywhere the most abundant. If they resembled their extant relatives, all three were grass eaters and were gregarious. Therefore, if there were big herds of large grazing animals, there must also have been enough forage for them, and it must have been of an appropriate kind. Ordinary tundra vegetation as it exists today would not have been adequate.

Estimates have been made of the live-weight (*standing biomass*) per unit area of the large mammals living in two strongly contrasted modern ecosystems. In kilograms per hectare, the results are three or less for tundra and around thirty for prairie grasslands in Alberta, Montana, Wyoming, and South Dakota.[3] Paleoecologists who believe there were herds of big game therefore conclude that ice age Beringia must have been covered with highly productive vegetation differing markedly from modern tundra and containing plenty of grasses. They have named this vegetation *arctic steppe* (some prefer the term *steppe tundra*). Other paleoecologists emphatically reject the notion that arctic steppe could ever have existed. Let us examine the evidence offered by both sides in the debate and try to visualize how Beringia must have looked, first assuming that the arctic steppe hypothesis is correct and then assuming it is false.

Arctic steppe is assumed to have been a form of vegetation combining the qualities of modern steppe (the Russian word for midlatitude grasslands) and arctic tundra. After grass-eating mammoth, horse, and bison, the next commonest species were caribou and muskox, both typical tundra dwellers. Among the less abundant animals were several other grass eaters, for example, wapiti, camel, and saiga. The chief argument in favor of the arctic steppe hypothesis is that it must have existed if herds of all these animals were able to survive in the region.

Although most paleoecologists agree that the Beringian climate was probably much drier than at present, they disagree about temperatures. There are good arguments for a cold climate (see The South Coast of Beringia, chapter 6), but some paleoecologists[4] believe that summers may have been warmer than they are now. Fossil remains of badgers have been found at Dawson in the Yukon, but nowadays they do not live north of the warm midlatitude grasslands south of the boreal forest.

Badger

5 cm

Lambe's Horse

10 cm

Conditions were unsuitable for trees except for occasional groves of balsam poplar in sheltered spots. A 25,000-year-old fossil poplar log was found at the mouth of the Colville River in Alaska on the shore of the Arctic Ocean (fig. 7.1), and poplars still thrive in the region today. It seems likely that they held their own in this arctic setting throughout the whole of the Wisconsin glaciation.[5] Apart from occasional poplar groves, however, the landscape was treeless and, like modern prairies and steppes, windswept. Tree growth was impossible in exposed sites for three reasons: water shortage, abrasive sand storms, and constant trampling by the hooves of grazing animals that would have destroyed tree seedlings as soon as they germinated. Conversely, it was the lack of trees that enabled the relentless gales to blow unimpeded. Evidence of fierce winds remains to the present day: tracts covered with thick layers of *loess* (a dust of clay particles, silt, and fine sand deposited by the wind), and large sand dunes are scattered throughout modern Alaska and the Yukon.

The vegetation over vast areas is assumed to have been grassland (the arctic steppe). Studies of pollen and larger fossils show that it included at least one kind, grama grass, whose modern range is several thousand kilometers to the south.[6] There was also plenty of sage, probably pasture sage; in particularly arid places, there may have been more sage than grass.[7] A few fossils of other plants more reminiscent of today's prairies than today's tundra have also been found, for instance, wild blue flax and fairy candelabra.[8] But these species are still to be found growing in dry places in Alaska and the Yukon at a time when the land is *not* covered by widespread arctic steppe; there seems no reason to suppose that were any commoner in the past than they are at present.

So much for the arctic steppe scenario. Now consider the other side of the debate. According to many paleoecologists, there is no need to hypothesize anything so improbable as luxuriant grassland in arctic and subarctic Beringia at the peak of the last glaciation because the herds of huge mammals in search of forage were not nearly so large as they have been made out to be by defenders of the arctic

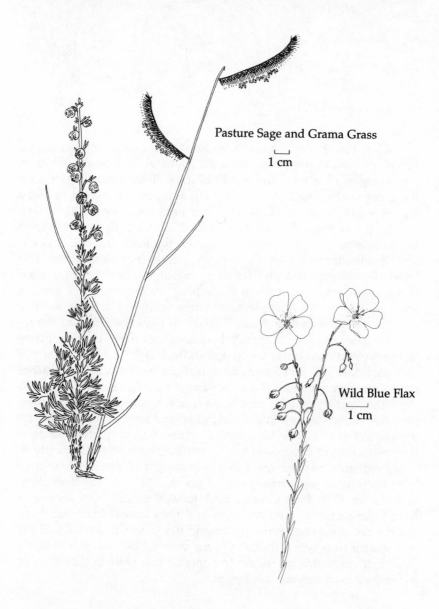

Pasture Sage and Grama Grass

1 cm

Wild Blue Flax

1 cm

steppe theory. The majority of the fossils that gave rise to the theory were found either in placer gold mines or river shallows, where they had presumably been redeposited by flowing water. They came from large areas of country and had been accumulating for a long time, perhaps several millennia,[9] in these "natural cemeteries." Therefore, there is no need to suppose there were great tracts of good pasture for huge grazing herds, but only that there were a few grassy meadows in a few valley bottoms, where small herds of animals could maintain themselves.

Another argument against the arctic steppe theory is that pollen samples have failed to provide any evidence for it. On the contrary, measurements made in the northern Yukon show pollen influx rates of only 100 grains per square centimeter per year. This is the amount of pollen that very sparse tundra vegetation, with much bare ground, would be expected to give; a dense sward of grass would give ten times as much.[10]

Human Life in Beringia

In addition to its other large mammals, Beringia also had a human population. The best evidence comes from fossils of other mammals found in river sandbars near Old Crow and in the Bluefish Caves, both in the Yukon (see fig. 7.1).

The signs of human presence consist of bones of various species of large mammals that have either been "worked" (fractured, flaked, ground, or polished) to convert them into tools or that show signs of having been "worked on," in the sense that they have been cut, scraped, or pierced as the animals they belonged to were butchered. As well, tools made from mammoth or mastodon tusk ivory have been found at Old Crow, and stone tools have been found at Bluefish Caves.[11] Old Crow in particular is famed for the tens of thousands of fossilized bones, teeth, tusks, and antlers that have been collected there; they have been subjected to detailed study under the leadership of the paleontologist C. R. Harington of the National Museums of Canada. There have been finds in Alaska, too, for example, at Fairbanks and Lost Chicken.

Radiocarbon dating shows that Beringia had a human population at least as early as 50k B.P. and has probably been continuously occupied ever since. At some places there are long gaps in the archaeological record; for instance, no humans appear to have been in the Old Crow district between about 21k and 14k B.P., but their absence from that locality was merely because during that period Old

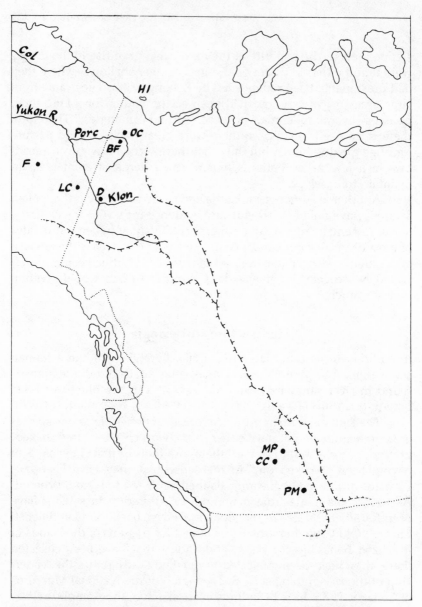

FIGURE 7.1: Map showing the places named in chapter 7. Also shown, by the symbols TTTTT, are those parts of the ice sheet margins forming the "walls" of the ice-free corridor. The corridor was temporarily blocked where the "walls" came together. The dotted lines are the international boundaries. Abbreviations: Col., Colville River; Porc., Porcupine River; Klon., Klondike River; HI, Herschel Island; OC, Old Crow; BF, Bluefish Cave; F, Fairbanks; LC, Lost Chicken; D, Dawson; MP, Mountain Park; CC, Castleguard Cave: PM, Plateau Mountain.

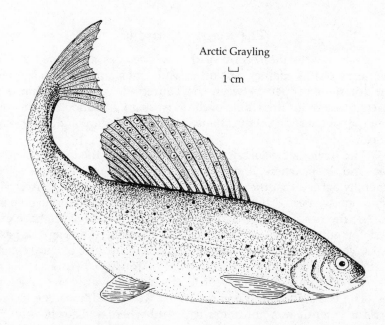

Arctic Grayling

⊔
1 cm

Crow was submerged by a proglacial lake (see chapter 1) that formed where the Porcupine River was dammed by the Laurentide ice sheet (for more on this and other proglacial lakes, see Migration from Beringia, chapter 9).

To survive in the harsh climate of Beringia requires skill. The people must have made the necessary clothes, shelter, and fire to keep warm through arctic winters and must have been capable of designing, constructing, and using workable tackle for hunting and fishing, or they would have starved. To hunt and fish successfully they would have had to travel, on land and water in summer and over ice and snow in winter.

It is difficult to judge whether food was abundant or scarce. As described in the previous section, paleoecologists disagree on whether the large mammals were present in vast numbers or only in small, local herds. But other quarry was available. Fossils of many different birds have been found: ducks, geese, loons, grebes, gulls, shore birds, and ptarmigan or grouse.[12] There were also fish, including northern pike[13] and arctic grayling.[14] One other animal was present, and it was not (we hope) an article of food: among the Old Crow fossils are remains of the domestic dog.[15] Beringia may well have been the place where dogs were first domesticated. Dog fossils have also been found on Herschel Island, which presumably means that humans were there too.

The Ice-free Corridor

We come now to the second of the two debates mentioned at the beginning of this chapter. The question is not simply, Did an ice-free corridor remain open between the Laurentide and Cordilleran ice sheets throughout the whole of the Wisconsin glaciation? There are, rather, a whole series of questions, some glaciological and some ecological.

The obvious glaciological questions are: Did the two ice sheets meet and, if so, where, and when, and for how long? It seems to be generally agreed that the corridor was not open at all times. According to one scenario,[16] the ice sheets coalesced as shown in the map in fig. 7.1; the resultant blockage of the corridor is believed to have existed for not more than 10,000 years at some time during the past 50,000 years. If this is true, then a corridor in the glaciological sense has existed, with only a short interruption, for 40,000 years.

However, a glaciological corridor is not necessarily an ecological one. The ecological questions are: What were conditions like in the corridor when it was not under ice? And when, and for how long, was the corridor ecologically usable in the sense that it was habitable? Obviously the answer depends on what sort of inhabitants are being envisaged. Most of the research on these topics has been inspired by a wish to know when the corridor could have functioned as a route for humans spreading from Beringia into midlatitude America.

This problem is part of a much larger one. As noted in chapter 5, the first definite evidence of widespread human presence south of the glaciated regions dates from about 11.5k B.P. These people seem to have been culturally united in that they all made and used the distinctive Clovis-type spearheads. What is *not* known is whether these people were descended from earlier, pre-Clovis people living in the same region, about whose very existence there is considerable doubt, or whether they were the descendants of immigrants from Beringia. The problem can also be looked at from the Beringians' point of view. Did the Beringians die out leaving no successors? Or were they the ancestors of the much more advanced Clovis people?

Assuming that humans spread southward from Beringia through the corridor, the answers to these questions depend on when the corridor became ecologically open. For much of the time when it was glaciologically open, it must have been a forbidding place.[17] Incessant, strong, bitterly cold winds would have blown off the nearby ice sheets. The land would have consisted of a mixture of bogs, marshes, islands of stagnant ice, and rock-strewn barrens, with dangerous, torrential rivers and icy lakes. Such vegetation as managed to grow would have been sparse. Food, clothing, and firewood would have been impossible to come by. While it was like this the corridor could not have served as a home for humans, and a home rather than a migration route was what was needed. It is unlikely that people migrated in the sense of marching steadily forward. They did not know what sort of destination lay ahead. Instead of journeying southward deliberately, the growing population simply expanded its hunting grounds, in whatever direction expansion was possible.

Because of this, some anthropologists argue that the Beringians described in the previous section could not have been the ancestors of the Clovis people unless the corridor became habitable considerably before 11.5k B.P., and opinions differ about whether it did. But this argument overlooks the strong possibility that humans spread southward via coastal refugia as far as the ice-free shores of what are now Oregon and California (see chapter 6). Subsequently, some of them may have migrated inland to the grassy interior plains, adapting their life-styles in the process. From being shore dwellers, they would have become big game hunters.[18]

The big game itself could not, of course, adapt to such strongly contrasted environments; grazing animals, for example, could only journey through the corridor when it had been free of ice, and free of the influence of nearby ice, long enough to develop into grassland or grassy woodland.

The effect of successive openings and closings of the corridor on one of the big game species, namely bison, has been studied by the paleontologist R. D. Guthrie of the University of Alaska.[19] Bison, as a group, have evolved rapidly, giving a series of well-differentiated species, and numerous fossils have been recovered. This enabled Guthrie to work out a highly probable scenario to explain their history. (It is not the only scenario; some equally believable alternatives have been devised.)

Bison invaded North America from an original home in Asia. The parental species, now extinct, is referred to as steppe bison, *Bison priscus* (an older name for the same species is *Bison crassicornis*). Herds of steppe bison found their way into eastern Beringia (now Alaska and the Yukon), an area that functioned like a lock in a canal. Bering Strait formed the upstream lock gate: it was open when sea level was low (the strait was replaced by the land bridge); it was closed when sea level was high (the land bridge was submerged). The ice-free corridor formed the downstream lock gate. These "gates" opened and closed alternately. When the glaciations were at their height, the land bridge from Asia into eastern Beringia was open, but the corridor was blocked by ice. During interglacials, the land bridge was "closed" (submerged by Bering Strait), but the corridor was open, linking eastern Beringia with the Great Plains of midlatitude North America.

At the time of the last glacial maximum the land bridge was open and the corridor closed (fig. 7.2). Two species of bison (both now extinct) lived in what is now North America: *Bison priscus* in Beringia, where they were merely part of the great Eurasian population; and *B. antiquus* south of the ice sheets in the Great Plains. *B. priscus*, the larger, had two shoulder humps and long, curved horns. *B. antiquus* had a single shoulder hump and straight horns; it was

FIGURE 7.2: Two bison species at 18k B.P. North of the ice *Bison priscus* is immigrating from Asia; south of the ice in the Great Plains is *Bison antiquus*, which has evolved in isolation from the *priscus* parent stock.

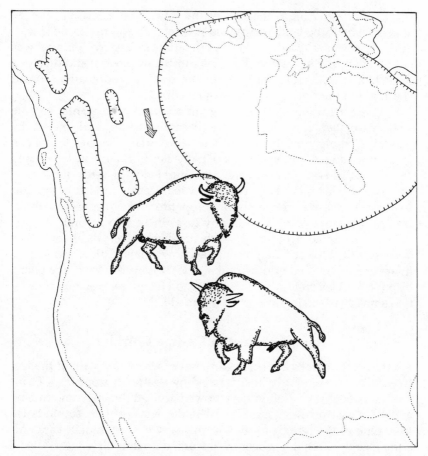

FIGURE 7.3: Two bison species at 10k B.P. Note that the map is on a larger scale than that in fig. 7.2. *Bison occidentalis* (with curved horns) has evolved from *priscus* ancestors and is traveling through the corridor to mingle with straight-horned *antiquus*.

descended from an earlier wave of *priscus* invaders, which had migrated through the corridor in the interglacial preceding the Wisconsin glaciation, known as the Sangamon interglacial. To complete the story, it is worth noting that the largest of all the extinct bison, giant bison (*Bison latifrons*), is thought to have descended from a wave of *priscus* immigrants that came through the corridor during the Yarmouth Interglacial, that is, the third interglacial counting back from our present interglacial as number one.

Then (fig. 7.3), at the end of the Wisconsin glaciation, the ice sheets waned, the land bridge was submerged, and the corridor widened. Consequently, the *priscus* population in eastern Beringia was cut off from that in western Beringia (northeast Siberia) and diverged from it in the evolutionary sense; the evolving population spread south along the now ecologically open corridor.

As generation succeeded generation, the migrants became slightly smaller, the curved horns became shorter, and one of the shoulder humps disappeared. The descendant species has been given the name *Bison occidentalis*. Finally, herds of *occidentalis* reached the Great Plains, where they met and mingled with herds of *antiquus*; the two species must have competed for pasture, and it is thought that *occidentalis* was the superior competitor. They probably hybridized to some extent as well. In any case, these two separate species were soon replaced by one, which has been given the name *Bison bison*; it is the bison of the present day, now almost extinct. Bison now living are subdivided into two subspecies: *Bison bison bison*, the plains bison; and *Bison bison athabascae*, the wood bison, represented now by two small herds living in northern Canada.

Refugia Near the Ice-free Corridor

Numerous nunataks are believed to have existed just west of the ice-free corridor. Valley glaciers descending eastward from the Cordilleran ice sheet would often have flowed around the lower mountains and hills east of the Continental Divide, leaving their summits ice free. One of the refugia formed in this way is at Mountain Park, Alberta (see fig. 7.1); it has been well studied botanically.[20]

This refugium is the site of twelve disjunct plant species, of which four are mosses and eight flowering plants. The present geographical ranges of two of the latter—a species of cotton grass (*Eriophorum callitrix*), and skunkweed—are shown in figure 7.4. The cotton grass is an arctic-alpine plant, widespread in the arctic tundra but with only a limited distribution on the eastern slopes of the Rockies. Skunkweed is an alpine plant whose northernmost outpost is at Mountain Park; most of its range is in mountains to the south. It is exceedingly unlikely that as many as twelve different species, each growing at a considerable distance from the Mountain Park site, would all have been blown in to the same isolated "island" of disjunction after the ice sheets melted, especially as none of the species is adapted to spread over long distances. None of the flowering plants in the group has seeds with wings or parachutes, or seeds embedded in fruits attractive to birds.

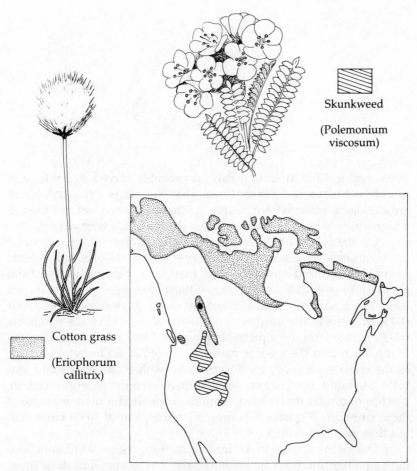

Skunkweed
(Polemonium
viscosum)

Cotton grass
(Eriophorum
callitrix)

FIGURE 7.4: The present geographical ranges of two plants whose disjunct ranges overlap at the site shown by the black dot, Mountain Park, Alberta.

The only reasonable explanation of their presence at Mountain Park is that they are survivors. Probably they all had larger ranges, with much more overlap, when the ice-free corridor formed a strip of arctic terrain between the ice sheets; the tundra of the corridor, as well as that on the Mountain Park nunatak, would have provided them with suitable habitat. But when the climate warmed and the lowlands of the erstwhile corridor were taken over by warmth-loving forests and grasslands, their shared refugium shrank to its present small size.

Plateau Mountain in Alberta is a similar refugium near the ice-

Brown lemming

5 cm

free corridor. Like Mountain Park, it probably served as a refugium in different ways at different times, that is, first as a nunatak (there are geological reasons[21] for believing it to have been one) and then as a high-altitude island of tundra when the lowlands warmed up.

On the slopes of Plateau Mountain is a limestone cave containing thousands of fossil teeth and bone fragments of small rodents (lemmings, ground squirrels, and marmots), ranging in age from about 1k to over 20k B.P.[22] Many of these are remains of the brown lemming, a northern animal whose modern range stops short about 600 kilometers to the northwest of the fossil site. When the ice sheets were big, the lemmings probably ranged widely around the margins of the ice and in the ice-free corridor, as well as on nunataks. Then, as the climate warmed, the tundra zone with all its plants and animals migrated northward, leaving behind only a few shrinking patches of tundra on isolated summits. Plateau Mountain was one of these summits; it retained its brown lemmings until 1,000 years ago, but they are now extinct there.

One of the refugia in the ice-free corridor region is still functioning as a refugium. It is Castleguard Cave, a system of limestone caves in the mountains (fig. 7.1). Part of it is overlain by meadows and part by the Columbia Icefield. Some small crustaceans live in pools in the caves and can truly be said to be experiencing an existing ice age and surviving it in a subglacial refugium.[23]

Most cave-dwelling organisms (*troglobites*) are aquatic animals such as fishes, salamanders, insects, crustaceans, and snails that are specially adapted to living in permanent darkness. They are usually colorless and blind, either because their eyes are degenerate or because they have no eyes at all. Troglobites are common in caves in regions that have been ice free throughout the present glacial age, but they are exceedingly scarce in formerly glaciated regions. This is believed to be because cave roofs often collapsed when ice sheets advanced over them; pools on the cave floors were scoured by moving ice, and no living things survived. And even if a cave was not destroyed, the water in it would usually have been frozen for centuries. The geological evidence shows, however, that Castleguard Cave has been unfrozen for more than 700,000 years.[24] The crustaceans now in the cave are probably the living representatives of populations that have survived there for at least that long.

There are two species. One is an isopod, *Salmasellus steganothrix*, which looks like a small (eight millimeters or less), delicate, colorless, blind sowbug; the other is an amphipod, *Stygobromus canadensis*, which looks like a small (five millimeters or less), delicate, colorless, blind beach flea. The amphipod is unique to Castleguard Cave, where it has been found only in one small group of pools; so far as is known, it occurs nowhere else in the world. The isopod is a common animal; it is quite abundant throughout the Castleguard Cave system and has also been found in water from other nearby cave systems in the eastern slopes of the Rockies in Alberta.[25] This is a disjunct part of its range, however; the main part lies a long way to the south, in unglaciated North America, at what is almost certainly too great a distance for it to have migrated into Alberta after the Wisconsin ice retreated.

Thus neither species could have invaded the cave after the last glacial maximum; they must have been there during the glaciation. The only problem is: During millennia of subglacial life, what did they eat? The animals have low metabolic rates, and their modest food requirements could no doubt have been easily met from two sources. First, subglacial streams percolating down into the cave presumably bear organic detritus that has been in rock crevices under the ice since before the current glaciation that is still not all washed away. Second, microscopic unicellular algae living on the surface of the snow are carried down into subglacial, and then into subterranean, streams; the most abundant of them is *Chlamydomonas nivalis*, the species that causes the "pink snow" so familiar to winter travelers in the mountains. No doubt these algae serve as fodder for the crustaceans today, and did so in the past.

PART THREE

THE MELTING OF THE ICE

8

The Ice Begins to Melt

Not long after 20k B.P. the climate began to warm and the ice sheets to melt. At the outset, the melting reduced the thickness of the ice without much affecting its extent. The result was that for a few thousand years the area of ice-free North America contracted rather than expanded. This was because tremendous quantities of meltwater reached the sea, bringing about a rise in sea level and submerging large tracts of low-lying land; not until later was there a compensating shrinkage in the area of the ice.

The great melt proceeded slowly at first. Then, by about 14k B.P. it speeded up for a millennium or so. Temperatures may have increased for astronomical reasons; there was, as well, another cause for accelerated melting.[1] By about 14k B.P., the margins of the ice sheets had drawn back within the coastlines of the North American continent and could no longer calve off bergs into the sea; coastal seas were therefore no longer cooled by floating ice and began to warm up. Onshore winds were warmer than they had been, and the thinning ice sheets began to shrink in earnest.

Meltwater from the Laurentide ice sheet formed great freshwater proglacial lakes in the center of the continent; the lakes were trapped between rising land on one side and the ice front itself on the other, and could not flow away (see figs. 1.3 and 1.4, and Ice and Fresh Water, chapter 1). Their existence hastened the shrinkage of the ice sheet. This happened because bergs calved off by the ice cliffs forming the northern shores of the lakes drifted away into warmer, more southerly waters, where they melted. This loss of ice by the continual loss of bergs proceeded especially fast when lobes from the ice sheet surged rapidly into the lakes owing to a buildup of pressure from behind.[2]

As the margin of the active ice sheets retreated, four different kinds of terrain appeared at different places in their wake: open ground; tracts of stagnant, drift-covered ice; proglacial lakes; and extensions of the sea. Chapters 8, 9, and 10 describe how life established itself in these different environments. The process occupied the final millennia of the Pleistocene epoch.

South of the Ice: Tundra

As soon as a tract of land was ice free, the materials needed for vegetation to get started—soil and seeds—were brought in by the incessant strong winds that blew along the ice front unobstructed by vegetation. Drifting soil, blown in from unglaciated land, often provided fertile seed beds, and the winds also carried seeds of hardy, pioneer plants, perhaps from distant sources. The first vegetation to follow the ice was quickly established and began the well-known process of plant succession.

Consider events south of the ice. Places as far apart as Norwood near Minneapolis[3] and Athabasca in central Alberta[4] (fig. 8.1) were colonized by the same pioneer plant species as soon as the ice disappeared: a mixture of poplar and various shrubs, such as willow, juniper, buffalo berry, and silverberry (also called wolf willow).

These similar plant communities were not only far apart spatially, they were separated by a long time interval as well. In both places the pioneer vegetation was ultimately crowded out by advancing spruce forests; this happened at about 12k B.P. in Norwood but not until about 11k B.P. in Athabasca; at 12k B.P. Athabasca was still under ice. Events such as the arrival of spruces, which happened at different times in different places, are known as *time transgressive*.

The kind of vegetation to succeed the first communities of pioneer plants varied from place to place. For example, in southern Minnesota the pioneer vegetation was followed by spruce woods; in northern Minnesota, by tundra. The cause of the difference was presumably permafrost (see The Starting Conditions, chapter 4); only tundra could develop where the ground was in the grip of permafrost, and trees could not succeed tundra until the ground had thawed. Where tundra developed, it was not necessarily identical with the adjacent tundra on ground that the ice sheets had never reached. Studies of fossil pollen show that in never glaciated parts of Pennsylvania the tundra was dominated by grasses, whereas in recently glaciated Connecticut, it was dominated by sedges.[5]

The rate at which the permafrost thawed also controlled the rate at which the boundary (or *ecotone*) between the tundra zone and the

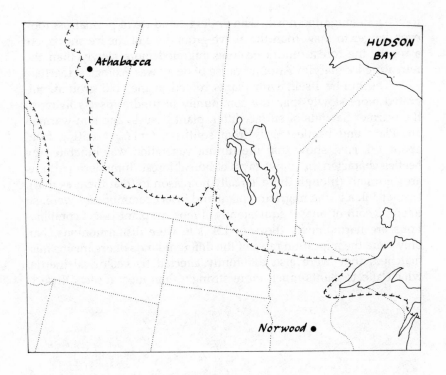

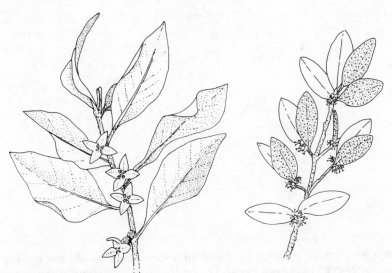

FIGURE 8.1: The location of Norwood, Minnesota, and Athabasca Alberta. The position of the ice front at 12k B.P. is shown by ⊢⊣⊢⊣⊢⊣ . Below: left, silverberry; right, buffalo berry.

forest zone migrated north. The underground ice of permafrost took even longer to thaw than the above-ground ice of the ice sheets. As a result, the forest-tundra ecotone migrated more slowly than the retreating ice margin. Another cause of delay was ecological inertia.

The tundra itself, with plants rooted in the cold ground, migrated more slowly than the community of tundra insects living in the warmer habitats of surface litter, plant tissues, and sun-warmed air. Plant and beetle fossils from southern Ontario,[6] dating from about 13k B.P., show that the tundra vegetation was inhabited by beetles characteristic (nowadays) of boreal forest. If modern patterns are "normal" (though there is really no reason to treat them as that), then at 13k B.P. the migrating plant and insect communities were, so to speak, out of phase; equivalently, if modern plant-insect combinations are harmonious, those of 13k B.P. were disharmonious. Not only were the migration rates of the different kinds of organisms mismatched; they were also differently affected by ecological inertia, which affects plants much more strongly than most insects. Insects

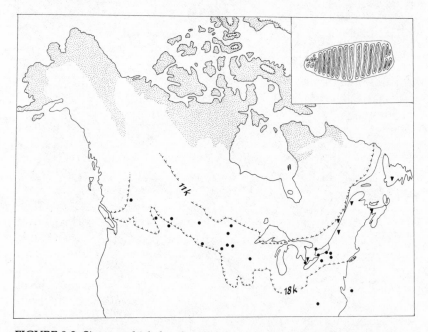

FIGURE 8.2: Sites at which fossil teeth of woolly mammoth (●) and fossil ice wedges (▼) have been found. The positions of the ice front at 18k and 11k B.P. are shown by ⁺⁺⌄⌄⌄⁺⁺. The stippled area is land at present covered with tundra. Inset, mammoth tooth, showing the biting surface.

that depended for survival on particular food plants, rather than making do with whatever was available, were exceptions, of course; they could not migrate faster than their fodder.

Turning from beetles to more conspicuous animals, consider mammoths. The map in fig. 8.2 shows sites where the cheek teeth of woolly mammoths have been found,[7] mostly between the ice front positions for 18k and 11k B.P. The same map shows the sites of fossil ice wedges[8] indicative of ancient permafrost. Both woolly mammoths and permafrost suggest tundra. But even though the symbols on the map appear to form a strip almost crossing the continent, few paleoecologists believe that the tundra zone was continuous. The symbols show a noticeable gap around the upper Great Lakes, where forest rather than tundra followed behind the melting ice.

South of the Ice: Forest Parkland and Muskeg

The first forests were probably not closed; more likely they consisted of an open parkland, or a mixture of forest and barrens, where conditions were to the liking of both grazing and browsing mammals. Mastodons were numerous; the region south of the Great Lakes is especially rich in mastodon fossils.[9] Open forest is better suited to browsers than closed; in closed forest the edible, foliage-laden twigs are too high above the ground for browsers to reach them. Open forest is also better suited to human hunters; a few scattered bands of hunters were probably around, after 11k B.P. if not before (see chapter 5). In the following discussion of these forests, open, well-lit, parklike forest should be envisaged.

The first forests to become established in the wake of the ice seem to have been unlike the modern boreal forest in many ways, even though spruce, mostly black spruce, was the most abundant tree species. Consider the forests of the Midwest, especially Minnesota, where very detailed studies have been made. The pollen spectra left by the earliest forests of the Midwest have no modern analogues. As now, there was plenty of poplar (including aspen) and some birch, but other hardwoods, in particular ash, elm, and oak, formed a much larger proportion of the trees than they do today.[10] And there were no pines; their absence would have been as great an oddity (to modern eyes) as the presence of numerous hardwoods. Presumably they were absent because they had not had time to reach the Midwest from their glacial refugia in the east (for more on pine migration, see chapters 4 and 14).

Black ash and white elm are still found in the modern boreal

Mixed Woodland

forest but only near its southern margin, north of the Great Lakes. Burr oak grows in an even smaller sector of the boreal forest, in southern Manitoba. As the boreal spruce forest with its northern hardwoods (poplar and birch) migrated north in the wake of the ice, these more southerly hardwoods were left behind. At first thought, one might have expected the first postglacial forests at midlatitudes, immediately south of the tundra or the ice, to resemble today's boreal forest at high latitudes, immediately south of the modern arctic tundra. But evidently it did not.

The most likely explanation is that the climate of immediately postglacial midwestern America has no modern analogue. The summers were probably as cold as those in northern Canada today, partly because climatic warming had not proceeded far and also because the huge Laurentide ice sheet was so close. The climate was also wet. Then, as now, cold, wet summers caused the development of muskeg.

The winters, however, must have been considerably warmer than they are in the far northern forests of today. No elms, ashes, or oaks can endure temperatures as low as those tolerated by black and white spruce, and their presence in the late Pleistocene spruce forests proves that the winters were mild enough for them. There are two entirely different forms of cold hardiness in trees.[11] Some of the hardy species survive temperatures down to minus forty degrees Celsius, but no lower, because the sap in their living cells can supercool to that temperature without freezing; at lower temperatures the sap forms ice crystals and the trees are killed. In other species the sap in cooled tissues escapes from the interiors of the living cells into the intercellular spaces, where it freezes without doing any damage; these very hardy species can endure temperatures lower than any that occur naturally. Only very hardy trees (black and white spruce, tamarack, jack pine, balsam fir, trembling aspen, balsam poplar, and paper birch) grow as far as the northern limit of trees at the present day.[12] Trees that rely on the supercooling process to withstand frost, among them ashes, oaks, and elms, cannot grow north of the isotherm for minus forty degrees Celsius minimum winter temperature, which lies far south of the present northern limit of trees.

It follows that winters in the late Pleistocene forests of the Midwest must have been warmer than winters in the forests of far northern Canada today. This is not surprising. The lower latitude of the Midwest ensured more hours of daylight in the winter; also, the great domes of ice to the north blocked the incursions of cold arctic air masses so familiar in modern winters.[13] However, besides the relatively tender ashes, oaks, and elms, midwestern forests also contained some typically arctic ground plants now found only in arctic or alpine tundra. Examples are the spikemoss *Selaginella selaginoides* and a saxifrage that was probably purple saxifrage. Spores from the spikemoss and pollen from the saxifrage have been found in lake sediments dating from about 12k B.P. in northern Minnesota.[14]

As noted in chapter 4, the northward march of the forests seems to have been uninterrupted, in spite of local readvances of the ice sheet from time to time. The readvances are inferred from geological evidence: layers of glacial till of different ages superimposed on one another, terminal (ice front) moraines deposited at a succession of places as the ice melted, ancient beaches marking the shores of ice-dammed lakes, and the like. This geological evidence does not show whether a readvance was due to climatic cooling or to surging.

One of the big readvances of the ice in the Great Lakes region happened around 13k B.P., during what is known as the Port Huron *stade*.[15] (A stade is a cold spell within a glaciation; the relatively mild

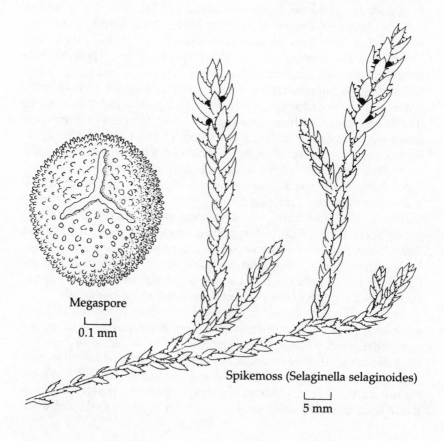

Megaspore

L———J
0.1 mm

Spikemoss (Selaginella selaginoides)

L———J
5 mm

spell between two stades is an *interstade*.) To call the period a stade is to imply that the readvance was due to climatic cooling. Other evidence, however, suggests a surge of ice. The surge split into four lobes that flowed into the almost ice-free basins of Lakes Superior, Michigan, Huron, and Erie-Ontario, respectively, while ice to east and west of the lobes continued its retreat. This certainly suggests surging, as does the lobes' speed of flow. They advanced at about two kilometers per year, which is very fast compared with the rate of advance of the margin of a growing ice sheet.

The absence of paleoecological evidence for climatic cooling does not necessarily mean there was no cooling, since ecological inertia could have damped the vegetation's response. Even so, it makes the surge theory that much more likely. Where the surging ice advanced over forested land, it must have crushed the full-grown forest ahead of it. And when the surging stopped, and the ice cliffs melted back, it would have been forest, rather than tundra, that repossessed the land.

Stagnant Ice

One of the most interesting postglacial terrains, in its strangeness to modern eyes, was stagnant ice. The retreating southwestern margin of the Laurentide ice sheet left extensive areas of stagnant ice in its wake, particularly in Saskatchewan, North Dakota, and Minnesota. There was some, too, at the southern edge of the Cordilleran ice sheet in the region of Puget Sound.[16] Stagnant ice, with fully developed forest growing in unfrozen soil on top of it, is found in North America nowadays only in a few places in the Alaska panhandle, where large glaciers (for instance, the Malaspina and Fairweather glaciers) reach the ocean.

There would always have been zones of stagnant ice "left behind" by the margins of the shrinking ice sheets. The zones would have been wide if the climate were cold and the ice sheets were shrinking because of starvation; they would have been narrow if the climate were relatively warm and the ice sheets were shrinking because of melting *unless* the stagnant remnants happened to be insulated. In several places they were very well insulated by accumulated glacial drift, and the result was long-lasting tracts of stagnant ice terrain.

A well-studied example of such a tract is the Missouri Coteau region of North Dakota (fig. 8.3), where geological and paleoecological evidence shows that a wide (over sixty kilometers) zone of stagnant ice persisted from about 12k to 9k B.P.; it was as much as 100 meters thick to begin with. The way in which this ice came to be insulated was mentioned in chapter 1; it is believed that a readvance of the ice sheet deposited a layer of glacial till on top of "old" (and cold) stagnant ice left by an earlier retreat. When the "new" ice melted, the layer of till insulated the old ice beneath.[17]

Farther east, in Minnesota, much of the insulation consisted of glacial outwash, silt, and gravel carried away in torrents of meltwater flowing from the front of the retreating ice sheet. Sediment cores from several lakes show that they are underlain by outwash; others are underlain by till. Yet another of the ingredients of the insulating layer was wind-borne dust (loess).

The thick layers of various kinds of insulation overlying the stagnant ice provided ground ready for invasion by plants. For the most part, conditions were no different from those on newly ice-free land that was *not* underlain by stagnant ice, and pioneer vegetation

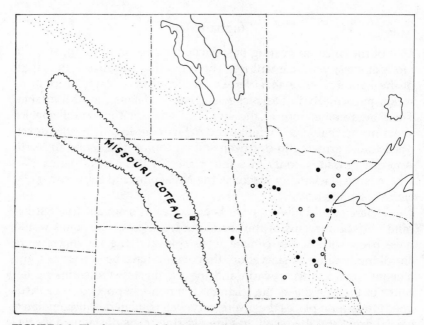

FIGURE 8.3: The location of the Missouri Coteau and of sites (chiefly in Minnesota) where drift from an early advance of the ice underlies lake sediments; ●, sites with wood debris below the lake sediments; ○, sites without; ☒, fossil fish site. The lightly stippled band shows, approximately, the modern transition zone between forest (to the east) and prairie grasslands (to the west).

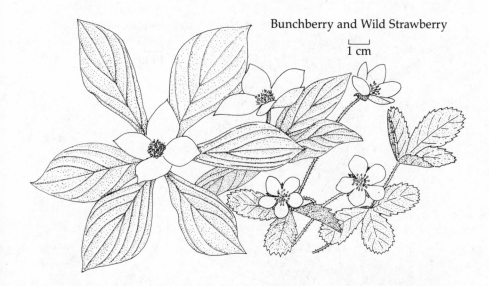

Bunchberry and Wild Strawberry

1 cm

followed by forest developed in the usual way. The development of forest in areas that today are covered by prairie grassland shows that the precipitation was higher then than now.

The evidence that the first vegetation was forest comes from sediment cores collected from lakes in the area. Between the inorganic drift at the bottom and the lowest (oldest) layer of lake floor sediment, there is sometimes (see fig. 8.3) a layer of terrestrial plant remains: needles, twigs, seeds and wood fragments of spruce and tamarack, and seeds of such forest floor herbs as bunchberry, wild strawberry, and violets.[18]

The presence, in many lakes, of forest floor debris below the oldest layer of lake sediment implies one of two things: that the lakes formed and filled on what had been forest floor or that the original lake floor consisted of a glacial drift and that the forest debris was carried in by streams. Paleoecological deduction has proved that the first of these possibilities is the more likely. Besides terrestrial plant macrofossils (seeds, twigs, and the like), the lowest sediments also contain diatoms of several species that live on soil, moss, or rocks but not in water. Moreover, the fragile valves of these diatoms are often undamaged and still fitted together in pairs. One example of such a species is *Melosira roseana*, shown in the drawing.

The discovery of these delicate diatoms, perfectly preserved among the woody debris, shows that the debris cannot have been swept roughly along in streams. It follows that the lakes must have come into existence spontaneously in the midst of rather ordinary

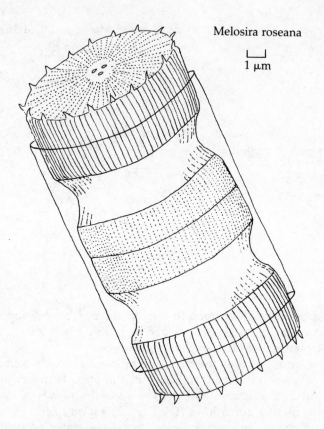

Melosira roseana

1 μm

forests. It was the development of these lakes that gave stagnant ice terrain its strangeness. The lakes merit a separate discussion.

Superglacial and Ice-walled Lakes and Their Ecology

The ice in stagnant ice terrain was for the most part buried deep, too deep for it to influence the environment on the surface. But this was not the case everywhere. There were occasional thin spots or even gaps in the insulating drift blanket, and here the ice began to melt. It melted in summer sun and dissolved in summer rain. Meltwater and rainwater together must have formed rivulets that drained into crevasses and then spread laterally, eroding a network of tunnels and caverns in previously insulated ice. Even while forest was in the process of developing over much of the area, the ground was crumbling here and there, where the stagnant ice had become honeycombed and weak around the thin spots and where cavern roofs caved in.

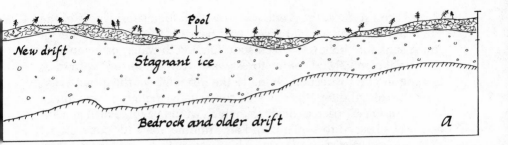

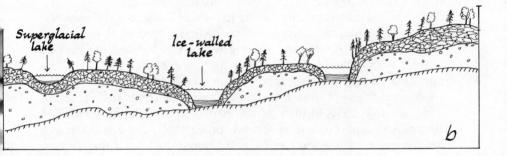

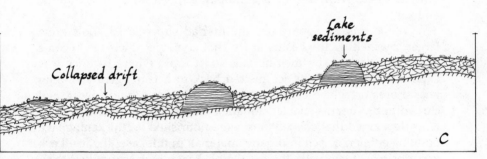

FIGURE 8.4: The sequence of events as beds of stagnant ice melted. *a*,
About 12k B.P. Small meltwater pools where drift is thin or absent; drunken
forest of scattered, stunted spruces. *b*, After 1000 or 2000 years. Ice thinner
with some gaps; overlying drift is thicker and forests more luxuriant; sediments accumulating in superglacial and ice-walled lakes. *c*, Today. Ice gone;
surface drift collapsed onto underlying earlier drift and bedrock; lakes
drained and their sediments left as flat-topped hills. The climate is arid, and
grassland has succeeded forest.

At the surface, the scene was one of sudden subsidences of the ground where the ice had given way beneath (fig. 8.4a). Patches of forest sank into hollows, which soon filled with water, drowning the terrestrial plants. Much of the forest was "drunken forest," with trees leaning in every direction owing to the instability of the ground. On a small scale, the topography was continually changing. Sometimes the slumping exposed a cliff of ice, which would begin to melt as soon as it was exposed to sunlight and air. Innumerable little superglacial puddles and pools were formed because of the subsidences. They were icy cold and probably, like glacial lakes today, milky with rock flour; as environments for life, they were unpromising.

The continued melting of the ice where it was poorly insulated improved insulation. The negative feedback process that allowed melting ice to slow its own melting worked as follows: the ice was "dirty," with embedded rocks (how dirty is not known), and as the surface gradually sank lower at the thin spots because of melting, the rocks were released from their icy matrix and accumulated. Thus, in time, the thin spots in the original insulation automatically repaired themselves, and the melting slowed considerably. The insulation was also augmented everywhere by wind-borne dust from the zone of newly exposed land next to the ice sheet margin (which was not far away to begin with) and by accumulating forest floor litter, humus, and soil.

While this was going on, the myriad superglacial pools grew larger, by melting the basins of ice that contained them, to become ponds and lakes. The melting was most rapid where bare ice was exposed. Steep banks of ice melted back to become gently sloping lake shores on which carpets of insulating debris could collect. Soil carried in by streams enabled aquatic plants to grow. Aquatic ecosystems developed that created, and were nourished by, thickening lake bottom sediments, rich in organic material; in this case, the feedback was positive. Eventually the insulating layers that everywhere covered the buried ice (soil over drift on the land, sediment over drift on the lake floors) thickened sufficiently for insulation to be complete. Life on the surface, both terrestrial and aquatic, was then entirely unaffected by the underlying ice (see fig. 8.4b). It is known from studies made in Alaska[19] that a drift layer two meters thick insulates surface lakes from buried ice completely.

Gradually the insulation became absolute. Gradually the terrain became less active in the sense that slumps and subsidences became less and less frequent. The melting of the buried ice (which continued to melt, of course, or it would be there still) came to be caused en-

Broad-leaved Arrowhead and Slender Naiad

1 cm

tirely by the earth's internal heat. And meltwater from the buried ice contributed progressively less to the lakes until they were supplied entirely by precipitation.

This last change was the most important ecologically. It allowed the lake waters to become clear enough and warm enough for life. Fossils of many kinds have been discovered that show conditions resembled those in the modern boreal forest rather than those in the modern arctic, in spite of the buried ice not far below the surface. The earliest aquatic plants, which have left seeds and pollen as evidence of their presence, include slender naiad, broad-leaved arrowhead, cattails, and pioneer species of sedges and bur reeds.[20]

Fossil diatoms, fossil ostracods, gyrogonites (fossil nucules of stonewort; see Microfossils, chapter 2), and fossil mollusc shells also aid in reconstructing the paleolandscape. The contrast with the modern landscape is especially noteworthy in the western half of the stagnant ice region, which includes the Missouri Coteau. Where the climate is now arid, it was then comparatively wet; where prairie grasslands now exist with an abundance of small alkaline lakes (the prairie potholes), spruce and tamarack forests existed then from about 12k to 10k B.P. with an abundance of small sweet water lakes.

It seems likely, though there is no direct evidence, that then, as

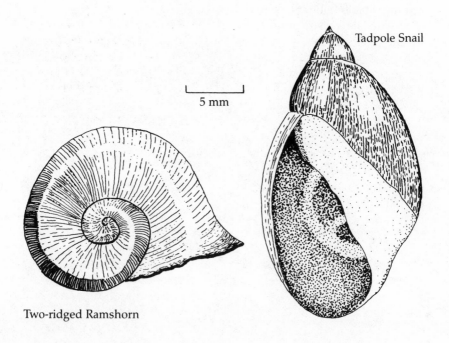

Tadpole Snail

5 mm

Two-ridged Ramshorn

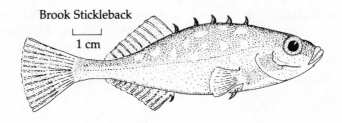

Brook Stickleback

1 cm

now, the lakes were stopping places for many species of migrating ducks and shore birds. Birds were probably responsible for introducing much of the life that first colonized the superglacial lakes. The birds' flyways cannot, of course, have been the same then as now; perhaps some species that now nest in the arctic nested then in the stagnant ice terrain; others may have continued north, around the Laurentide ice, to breed in Beringia and ice-free parts of the arctic farther east (see fig. 1.4). But whatever their destinations, migrating waterfowl and shore birds were probably plentiful and served as passive transporters of much other life; mud adhering to their feet would have contained a variety of microorganisms and also the seeds of aquatic plants and the eggs of aquatic snails, for example, those of the two-ridged ramshorn and the tadpole snail, which are both present in suitable habitats in the region today.[21]

The lakes also contained mussels of three species still common in the region.[22] Bivalves (clams and mussels), unlike gastropods (snails), cannot migrate in the absence of fish; their immature forms (glochidia) are parasitic on fish (see chapter 3), and the mussels known to have invaded the superglacial lakes must have come as passengers on fish of suitable species through rivers linking the lakes to tributaries of the Missouri River. Fish fossils dating from about 10k B.P. have been found at a site on the Missouri Coteau (see fig. 8.3), and four of them—brook stickleback, brassy minnow, blacknose shiner, and banded killifish—typically live in shallow ponds; superglacial ponds and ice-walled lakes probably provided excellent habitats for them.[23]

Stagnant ice terrain finally ceased to exist when the last of the buried ice melted, about 9k B.P. At first, lakes that had been superglacial were let down onto bedrock, becoming so-called ice-walled lakes for as long as the surrounding ice remained (fig. 8.4b). The last blocks to melt left the holes occupied by modern prairie potholes. All the land surfaces gently collapsed, because of the disappearance of the ice that had supported them. The lowered land levels allowed the ice-walled lakes to drain, and their sediments, which had accumulated layer upon layer to a considerable thickness, were left high and dry as flat-topped hills (fig. 8.4c). By the end of the melting period, at about 9k B.P., a reversal of the original topography had taken place, so that the fossils of lake-dwelling organisms are found on what are now hill tops.[24]

9

The Great Proglacial Lakes

The melting of the great ice sheets naturally produced enormous volumes of meltwater; the meltwater formed extensive *proglacial* lakes dammed on one side by cliffs of ice (see chapter 1). Many of the lakes were vast, and all of them changed their shapes and positions continuously as the ice front itself shifted. These great freshwater lakes were second in importance only to the ice sheets themselves as components of what may be called the geographical scenery (the view as it would have appeared from a satellite) during the late Pleistocene and early Holocene. The period of the great proglacial lakes lasted from about 15k B.P. when the first, comparatively small lakes began to form, until about 8k B.P., by which time, although large remnants of the ice sheets still existed, their margins no longer dammed any sizable lakes.

The lakes were of tremendous ecological importance in a number of ways. The aquatic ecosystems that developed in them were as much a part of the postglacial biosphere as the terrestrial ecosystems on dry land; the lakes also influenced the neighboring terrestrial systems by forming barriers to migrating land plants and animals for part of the time and then, when a lake shifted its position or drained away, by leaving a tract of newly exposed lake mud ready for colonization. The presence of huge bodies of water adjacent to the ice affected the climate. The lakes were also junction points in a drainage network whose pattern changed repeatedly as the ice sheets wasted away; for example (see Glacial Lakes Agassiz and McConnell in this chapter), rivers that had drained to the Gulf of Mexico at glacial maximum subsequently changed course and drained first into the Arctic Ocean and then into the Atlantic. Lastly, some of the rivers flowing out of the lakes came into existence abruptly and spectacularly in a way that has no parallel in the modern world (see following section).

This chapter outlines the history of these great proglacial lakes over the 7,000 years (approximately) of their existence. The time interval covered crosses the Pleistocene/Holocene boundary, which most geologists put at 10k B.P. This date is not arbitrary, in spite of its being such a satisfyingly (and suspiciously) round number. The millennium centered on 10k B.P. was marked by rapid and profound changes in the terrestrial biosphere, as we shall see in chapters 11 and 12, making the date a useful boundary point in the paleoecological history of the land. But there was no correspondingly abrupt change in the progress of the lakes.

Progress is the appropriate word: the earliest, comparatively small lakes formed in the west, south of the Cordilleran ice sheet; they were followed by lakes on the western and southwestern sides of the Laurentide ice sheet. With the passage of time, as Laurentide ice—the dam that retained them—wasted away, these latter lakes migrated eastward, growing enormously. To follow the story, therefore, we shall proceed forward through time and simultaneously forward (from west to east) through space.

Glacial Lakes Missoula and Columbia

The two greatest western lakes, Glacial Lake Missoula and Glacial Lake Columbia, formed immediately to the south of the Cordilleran ice sheet some time before 15k B.P.[1] (fig. 9.1). As already noted, they were comparatively small as proglacial lakes go, but this does not mean they were small by modern standards. At its maximum, Lake Missoula was about 300 kilometers long, and its volume was about equal to that of modern Lake Ontario, somewhat over 2,000 cubic kilometers.

What makes Lake Missoula remarkable is that it alternately filled and emptied in a cycle that was repeated more than forty times in a period of perhaps 1,500 years. The cycle operated as follows.[3]

The lake was contained by rising ground to the southwest and by the ice sheet to the northeast. There was also, at its extreme western end, a relatively narrow ice dam formed by a southward-pointing tongue of ice known as the Purcell trench lobe; the tip of the lobe filled the valley now occupied by Lake Pend Oreille in northern Idaho. It was the behavior of this lobe that governed the lake's cycles. Let us trace one cycle from the moment when the lake had last emptied.

The climate was gradually warming, and the ice sheet steadily melting. Therefore, accumulating meltwater slowly filled the lake basin. The process continued in orderly fashion for several years. Fi-

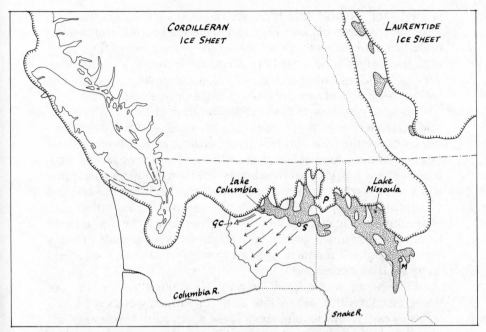

FIGURE 9.1: Glacial Lakes Columbia and Missoula south of the Cordilleran ice sheet at about 14k B.P.[2] The swarm of arrows mark the Channeled Scablands, across which floodwater flowed when a jokulhlaup emptied Lake Missoula. GC, Grand Coulee; P, Purcell Trench lobe; S, M, respectively, sites of the modern cities of Spokane, Washington, and Missoula, Montana.

nally, however, the lake became so deep that the ice dam at its western end began to float; the rising water lifted the buoyant ice clear of the lake floor, so that it no longer functioned as a dam, and the outcome was a catastrophic flood as Lake Missoula's waters suddenly flowed out under the ice, first into Lake Columbia and then in tremendous torrents that swept over the land and into the Columbia River valley. The whole lake probably drained away in less than two weeks.[4] Such a flood, caused by the failure of an ice dam, is known as a *jokulhlaup* (the Icelandic name, pronounced yo-kul-hape), or a *glacial outburst*. After each one, surging ice formed a new dam, Lake Missoula refilled, and the cycle repeated. There are believed to have been at least forty of these floods at intervals of between twenty and sixty years. They created the so-called channeled scablands of eastern Washington.

The evidence for these spectacular events is geological. There are channels where the floodwaters flowed out, of which the greatest is the Grand Coulee. And there are sediments left by the floods in lakes downstream of Lake Missoula. Fine-grained sediments accumulated on the floors of these lakes in the years between floods, but each succeeding flood from Lake Missoula deposited a bed of boulders, pebbles, and coarse gravel. As a result, alternating layers of fine and coarse sediments are now found in the downstream lakes.

It is also possible to discover the duration of each quiescent period between one flood and the next. The sediments laid down during a quiescent period formed clearly distinct annual layers, which can be counted. Annual layers of sediment (known as *varves*) accumulate in any still lake fed by streams that freeze in winter. In spring and summer, while the streams are flowing, they bring in sand and silt; for the rest of the year, while they (and the lake surface) are frozen, the only material settling on the lake floor is clay fine enough to remain suspended in the lake water for long periods. Thus a varve—one year's sediments—has a lower layer of sand and silt and an upper layer of clay.

The retreat of the Cordilleran ice sheet was thus the very opposite of uneventful, at any rate to begin with. It was marked by a series of cataclysmic jokulhlaups. These spectacular happenings did not take place in a lifeless world, and it is worth contemplating their ecological effects. The floods were obviously violently destructive, so the effects must have been negative. Each time Lake Missoula began to fill after one jokulhlaup, an aquatic ecosystem no doubt began to establish itself in the lake, but it could not have developed much before being entirely wiped out by the next jokulhlaup. The same must have been true of all the lakes and rivers that served as channels for the floods. Thus most bodies of fresh water in the region—in what is now the eastern half of the state of Washington—would have been noticeably lacking in aquatic life and probably differed conspicuously from waters that the floods bypassed.

The first jokulhlaups must also have swept away large quantities of terrestrial vegetation and the soil in which it was growing. The devastated region, the unattractively named channeled scablands, was initially covered with a thick deposit of loess. Pollen studies show[5] that it probably supported a patchwork of woodlands and grasslands. Sagebrush was abundant in the grassy patches; the woodlands contained spruces, firs, and pines. The pine pollen is believed to be that of species of five-needle pines; assuming that the climate was much cooler and wetter then than it is now, whitebark pine and western white pine are the most likely possibilities.

This was the land over which the jokulhlaups from Lake Missoula flowed periodically. The first floods eroded deep channels, which later floods followed. Huge quantities of soil, subsoil, and weathered rock were swept away, to be deposited downstream as outwash. The channel floors now have only thin soil and scanty vegetation over the basalt bedrock. Elongated "islands" of undamaged land, with the vegetation intact, were left between the runoff channels. Later, perhaps between 12k and 8k B.P., some tracts of the scoured land developed into patterned ground (see The Starting Conditions, chapter 4) because of frost action.[6] Fossil patterned ground is widespread in the area at present, especially in the northern half of the scablands, and there may well have been more in the past that has since been obliterated by the growth of trees.

Jokulhlaups continue to be fairly common occurrences in glaciated mountains, but modern jokulhlaups are orders of magnitude smaller than those of Glacial Lake Missoula. They also happen where glaciers reach the sea. Cases are known (for example, the Hubbard Glacier in Alaska) in which a surging glacier has closed the mouth of a fjord, turning it into a lake. Such a lake consists of pure sea water initially, but inflowing rivers gradually dilute it. After a time, the ice dam fails, and the lake and the sea outside suddenly become one again. If the dam lasts long enough for the lake to become fresh, both the marine ecosystem that was originally trapped, and later the freshwater ecosystem that developed in place of it, are destroyed. This sequence of events must have happened again and again on the west coast of British Columbia while the Cordilleran ice sheet was retreating. The ecological effects must have been devastating in individual fjords but would not have had widespread consequences for either marine or freshwater life.

Migration from Beringia

It was not easy for plants and animals to invade newly ice-free environments when the great ice sheets began to melt. Conditions were harsh and habitats inhospitable, at least to begin with. This was true for both terrestrial and aquatic organisms. Many aquatic organisms faced an additional difficulty: they required continuous water routes for their migrations. Not all kinds of aquatic life require water routes if they are to spread; for example, the seeds of water plants and the eggs of aquatic snails are unharmed by drying and can be carried in a viable state across dry land. But other kinds, especially fish, are completely unable to migrate from one drainage basin to another so long as the basins remain isolated. If a fish population is to spread, links must form, joining its "home" lakes and rivers to others.

The spread of aquatic organisms that had survived the Wisconsin glaciation in Beringia are good examples; they illustrate both the way migrations are begun and the way they are halted. At least six species of fish now living in the drainage basins of both the Yukon and Mackenzie rivers are believed to have survived the Wisconsin glaciation in the Yukon drainage, that is, in Beringia; among them are arctic grayling, northern pike, and lake whitefish. They evidently managed to migrate from their refugium in the Yukon drainage into the Mackenzie drainage, even though the two drainages are now separated by a height of land.

They are believed to have made the crossing in the following way.[7] At glacial maximum, the northern part of the Laurentide ice sheet had spread right across the valley of the Mackenzie River. When the ice started to melt, two proglacial lakes formed (fig. 9.2a), which filled and, in time, overflowed. The upper (southern), Lake Bonnet Plume, flowed into the lower (northern), Lake Old Crow, which flowed in its turn, via the Porcupine River, into the Yukon River. Aquatic organisms of the Yukon River drainage thus had access to the two proglacial lakes, and many migrated into them. Conditions were no doubt bleak. The land around the lakes was probably covered with scanty, high-arctic tundra, and the lakes would have been frozen for nine or ten months of the year.

This drainage pattern was only temporary. When the melting ice front had receded far enough to uncover the Mackenzie valley, the levels of both lakes fell and the channel linking them dried up. The lower lake (Old Crow) continued to empty westward, via the Porcupine. But the upper lake (Bonnet Plume) drained away eastward into the Mackenzie River; it ceased to exist as a lake and became the Peel River (fig. 9.2b).

This explains how species that had survived the Wisconsin glaciation in the Beringian refugium managed to invade the Peel River and its tributaries (which are now part of the Mackenzie drainage) when the ice sheets melted. But once they had reached it, they failed to spread very far. For example, the descendants of the lake whitefish stock that migrated into Lake Bonnet Plume from Beringia are now entirely confined to Margaret Lake (M in fig. 9.2b), a little lake in the valley of one of the Peel's tributaries.[8]

Most of the fish in the Mackenzie River today, except for those in its tributary the Peel, are descended from immigrants that came from the southeast, ultimately from the Mississippi drainage basin, in the southern, unglaciated half of the continent. We return in the following section to the topic of how they made the journey. Before that, two other questions remain. First, how is it known that several fish stocks currently living in the Mackenzie River are descended from

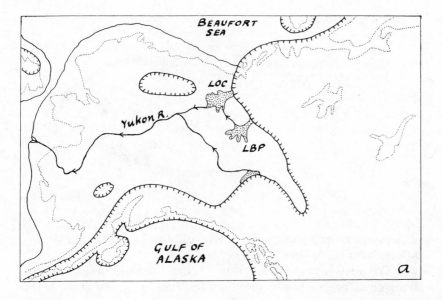

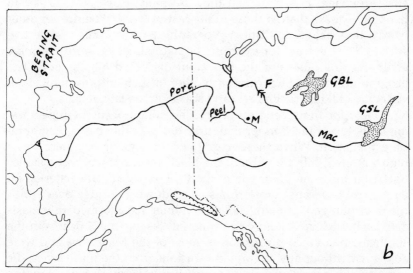

FIGURE 9.2: The drainage reversal that allowed fish of the Yukon drainage to reach the Mackenzie drainage. (*a*) The early stages of melting: LOC, Lake Old Crow; LBP, Lake Bonnet Plume. Together the lakes drained westward. (*b*) The modern drainage pattern: the Porcupine River, draining the Old Crow basin flows west; the Peel River, draining the Bonnet Plume basin, flows east. F, the site of the waterfall that blocked upstream migration (it has now disappeared). GBL, Great Bear Lake; GSL, Great Slave Lake; M, Margaret Lake; Mac, Mackenzie River; Porc, Porcupine River.

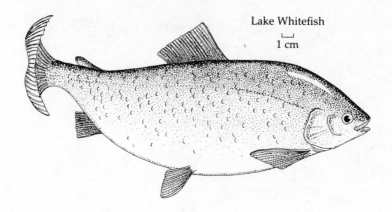

Lake Whitefish

1 cm

Mississippian ancestors? Second, why did the Beringian immigrants' descendants fail to spread farther?

The answer to the first question is that the fish stocks in the Peel drainage differ genetically from their relatives in the rest of the Mackenzie drainage. It is clear that they are much closer, genetically, to Beringian stocks than to those in the rest of North America. In other words, they belong to distinct geographical races of their respective species. This is true, for example, of lake whitefish, northern pike, arctic grayling, and also the lake trout discussed in chapter 3. The geographical races differ from each other in such traits as numbers of vertebrae, gill rakers, fin rays, and scales on the lateral lines.

Now for the second question. The failure of the Beringian immigrants to advance farther than they did probably has a geographical explanation. When these migrations were going on, there was a high waterfall (F in fig. 9.2*b*) in the Mackenzie, only a short distance upstream from the point where the Peel joined it. The waterfall no longer exists because erosion has worn down the ridge over which the river plunged, but while it lasted it must have formed an impassable barrier to fish.[9] Before the waterfall disappeared, fish from the southeast had taken possession of most of the Mackenzie and were able to crowd out any attempted invasions from the north when the route finally became passable, which it did about 6k B.P.. How the southerners arrived is the subject of the next section.

Glacial Lakes Agassiz and McConnell

The northward spread of fish—indeed of all aquatic organisms—from the unglaciated half of the continent south of the ice is bound up with the history of two great freshwater lakes: Glacial Lake Agassiz and Glacial Lake McConnell. Figure 9.3 shows them at their largest, at about 10k B.P..

The larger of the two, Lake Agassiz, came into existence about 12k B.P. and lasted for about 4,500 years. (It should be noted that experts disagree about these dates.) Its shape, size, and location changed continually (fig. 9.4) before it finally drained away, leaving as remnants only the modern Lakes Winnipeg, Manitoba, and Winnipegosis. It was the largest proglacial lake in North America; at its greatest, its area was about 350,000 square kilometers, which is more than four times the area of modern Lake Superior, now the largest freshwater lake in the world.[10]

The other giant lake, Lake McConnell, came into existence at about the same time as Lake Agassiz; like Lake Agassiz, it reached its greatest extent about 10k B.P. At its longest, 1,100 kilometers, it was longer than any freshwater lake in the modern world. In its early stages it was smaller, and the early version is known as Lake Peace. As it enlarged to its 10k B.P. maximum, it engulfed other ice front lakes. Then, as the ice front drew away northeastward, it separated again, this time into three "daughter" lakes that were left behind by the ice. These lakes were Great Bear Lake, Great Slave Lake, and Lake Athabasca, which have persisted more or less unchanged to the present day. So, in a sense, Lake McConnell still exists.

The two great lakes, McConnell and Agassiz, and the numerous lesser lakes that formed at the ice front, changed continuously in shape, size, and location. The changes were caused in part by the uncovering of new ground and the formation of new drainage patterns as the ice receded and in part by changes in the topography of the land itself, which underwent gradual isostatic adjustments as the weight of the overlying ice was removed. As the lakes changed, their drainages changed too. The arrows in figure 9.3 show some of the links that formed.

Both Lake McConnell and Lake Agassiz switched drainage directions several times over. Lake McConnell (or its smaller predecessor, Lake Peace) is believed to have drained first into Lake Agassiz, then into the Arctic Ocean via the Mackenzie River, then back into Lake Agassiz, and then back to the Arctic Ocean.[12]

Lake Agassiz, too, shifted its drainage direction four times.[13] Initially it drained south into the Mississippi valley. Then, at about 10.7k B.P., the ice front receded, opening up a spillway into the basin of the modern Lake Superior; Lake Agassiz's waters then flowed out to the east, and the level of the lake dropped. The old, southern spillway was left high and dry. Next, at about 10k B.P., lobes of the ice sheet surged southward in a temporary readvance that dammed the

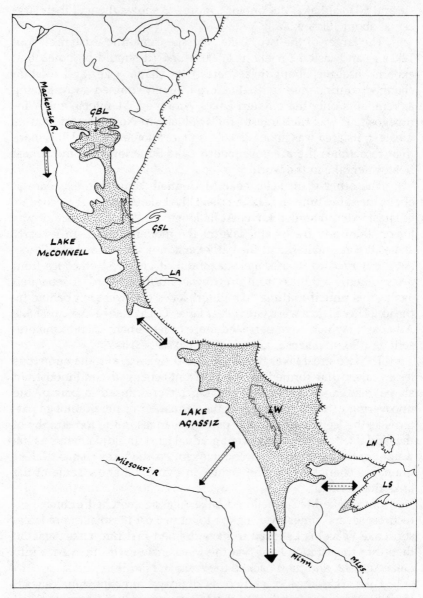

FIGURE 9.3: Glacial Lakes McConnell and Agassiz at 10k B.P.[11] The lightly outlined modern lakes are GBL, Great Bear Lake; GSL, Great Slave Lake; LA, Lake Athabasca; LN, Lake Nipigon; LS, Lake Superior; LW, Lake Winnipeg. The rivers draining Lake Agassiz are Minn, Minnesota River; Miss, Mississippi River. The double-headed arrows show where rivers formed at various times and through which fish could migrate.

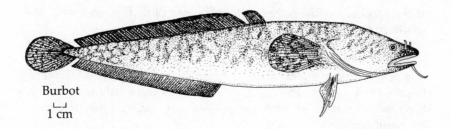

Burbot
⌐⌐
1 cm

eastern outlet once more; the level of the lake rose until it again over-flowed southward to the Mississippi. Finally (about 9.5k B.P.) the ice wasted away for the last time, opening a new outlet to the east; the new eastern outlet was some 500 kilometers farther north than the old, north of the modern Lake Nipigon (see fig. 9.3). At this time Lake Agassiz became joined to a younger, more easterly proglacial lake, Lake Ojibway (to be discussed in the following section), through which it drained away.

It seems likely that the majority of fishes living anywhere in the interior of Canada and the northern United States today have ancestors that, at one time or another, lived in Lake Agassiz. The lake was the hub of migration routes leading from the unglaciated southern half of the continent into most of the drainage basins once covered by the Laurentide ice sheet. Lake Agassiz received most of its immigrants from the south: first from the upper Missouri and its tributaries in the western Great Plains, to which it was linked by supergla-cial streams flowing across the tracts of stagnant ice southwest of the lake in the Missouri Coteau area; and later from the Mississippi valley, to which it was joined by the Minnesota River. Moreover, Lake Agassiz was sometimes linked, through Lake McConnell, to the Beringian refugium. As a result, fish species that survived the glaciation in refugia both to the south and to the northwest of the ice are now widely spread.

Two such wide-ranging species are burbot and northern pike. Both can withstand the cold of high latitudes. Burbot spawns in late winter, under lake ice; pike in the spring, when the ice is breaking up. Burbot is now spread throughout all the glaciated area except for parts of the Atlantic, Pacific, and Arctic coasts. Pike has nearly the same range, except that it is not found in British Columbia west of the Rockies. The two species survived in both refugia and invaded from both; however, stocks from the two regions differ genetically and it is known that fish of southern ancestry took possession of a far greater area than those from Beringia. The northwestern forms of burbot and pike are found only in Alaska and northern Canada west of the Mackenzie River; throughout the rest of their ranges, only the southern forms occur.

Today, there are 106 species of fish in the basin of glacial Lake Agassiz,[14] and all of them must have managed to migrate by at least one of the routes shown in the map in figure 9.3. Successful migrations were by no means assured. The rivers were only occasionally hospitable for fish. At other times they were raging torrents caused by jokulhlaups.

For example, between about 12k and 11k B.P. (before Lake McConnell grew big), a row of three ice front lakes existed west of Lake Agassiz that are believed to have drained eastward in domino fashion[15]; catastrophic overflow of the first sent huge volumes of water into the second, causing *it* to overflow catastrophically, and so

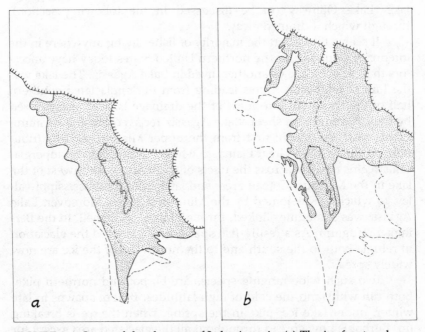

a   b

FIGURE 9.4: Some of the changes of Lake Agassiz. (*a*) The ice front and Lake Agassiz (stippled) at 10k B.P. The solid outline shows the modern Lake Winnipeg for reference. The dashed outline is that of Lake Agassiz at 12k B.P. (*b*) The ice front and Lake Agassiz (stippled) at 8.4k B.P. The long narrow lakes near its southwestern shore still exist; they are the modern lakes Winnipegosis and Manitoba (to north and south, respectively). The dashed outline is that of Lake Agassiz at 10k B.P.

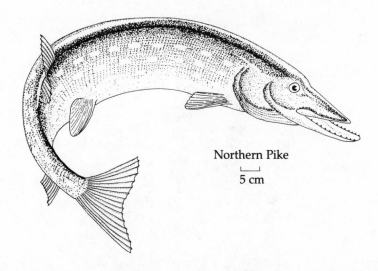

Northern Pike

⊢——⊣
5 cm

on. Great abandoned spillways (i.e., coulees) still remain (see the drawing in Ice and Fresh Water, chapter 1); they are believed to have been eroded by floods flowing at as much as 100,000 tons per second.

There were similar catastrophic floods when Lake Agassiz finally drained away eastward[16] at about 9.5k B.P. At one stage, the eastern shore of Lake Agassiz was separated from the western shore of Lake Nipigon (which was somewhat larger then than now) by a strip of land a scant ten kilometers across at its narrowest (see fig. 9.3). Numerous channels crossed the strip, draining Agassiz into Nipigon; they were the routes of spectacular torrents. A fire in 1980 burned away the forest and peat concealing the floor of the aptly named Roaring River channel and exposed a field of enormous boulders deposited by the floods; many were more than a meter in diameter.

The evidence for these dramatic events is geomorphological. Boulder fields show the routes, and the force, of torrential floods; old lake shores, river deltas, and lake bottom deposits, datable by the carbon 14 technique, show where proglacial lakes formed and how they migrated in the wake of the ice front. Ecological data enable us to visualize the settings in which these events occurred and discover their ecological consequences.

First, the settings. At present the region occupied by early Lake Agassiz in 12k B.P. is covered with prairie grassland. The vegetation that first took possession of the ground as the lake migrated northward was entirely different. It was probably an open woodland of

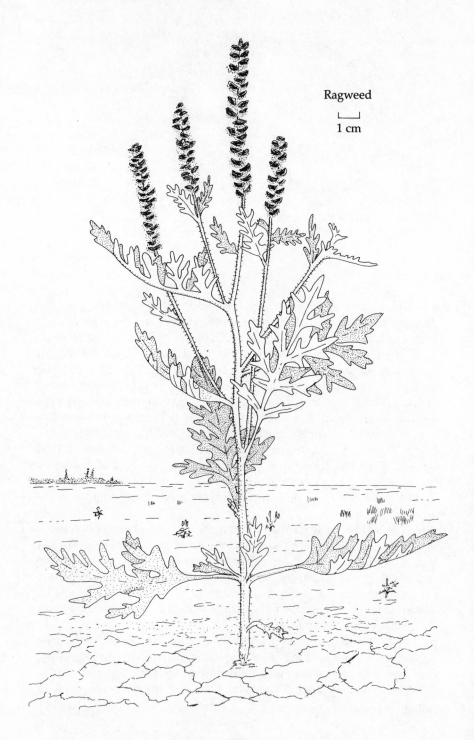

Ragweed

1 cm

spruce, poplar, paper birch, and ash, with sagebrush-covered clearings providing good habitat for woolly mammoths.[17] Subsequently, the spruce woods were replaced by pine woods, then by oak savanna, and finally by prairie[18] (these events are described below, in chapter 11).

The most unusual (to modern eyes) ecological phenomenon of the Lake Agassiz period was the abrupt appearance of an enormous expanse of freshly exposed lake bottom every time a glacial lake was suddenly drained by a jokulhlaup. Huge tracts of wet mud must have stretched as far as the eye could see (possibly there were a few human witnesses). As shown by pollen analysis, the contemporary "weeds" that invaded the newly available land were chiefly sagebrush and ragweed, with marsh vegetation in the wetter spots.

The Precursors of the Great Lakes and Glacial Lake Ojibway

The Great Lakes as they exist now form a single unit, draining into the Saint Lawrence River and out into the Atlantic. At glacial maximum there were no Great Lakes; their basin was completely covered by a lobe of the Laurentide ice sheet, the Michigan lobe. The precursors of the modern lakes were, of course, proglacial (ice front) lakes, and their geography changed continuously. The first two to form, at about 14k B.P., have been given the names Lake Chicago and Lake Maumee. Lake Chicago filled what is now the southern tip of Lake Michigan, and Lake Maumee was a larger version of modern Lake Erie; both drained south, into the Mississippi.

As the ice front retreated, more and more of the basin of modern Lake Michigan was uncovered (equivalently, Lake Chicago grew longer), and meltwater filled the basin of modern Lake Huron. These two bodies of water became joined and are known as Lake Algonquin (fig. 9.5). What had been Lake Maumee was left behind by the ice as a nonglacial lake, the precursor of Lake Erie; a new ice front lake, Lake Iroquois, more than filled the basin of modern Lake Ontario. These lakes drained into the Atlantic via the Hudson River.

Figure 9.5 shows the geography at about 12k B.P., at a time when terrestrial and aquatic ecosystems were rapidly diversifying. Tundra was giving way to spruce forests on the land south of the ice, and aquatic life began to flourish in the glacial lakes. Molluscs of many species became far more abundant than they had been in earlier glacial lakes;[19] there must have been algae and diatoms for them to feed on, and fish to support the clams' glochidia. The map in fig. 9.5 shows two fossil mollusc sites. The fossil collections differ slightly

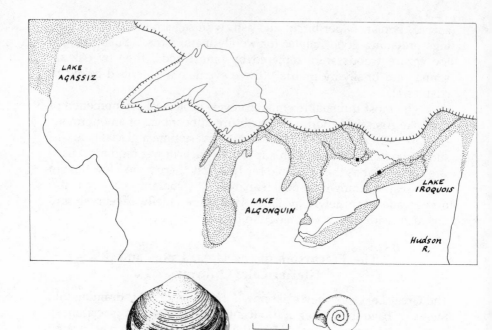

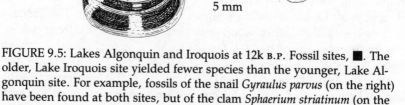

FIGURE 9.5: Lakes Algonquin and Iroquois at 12k B.P. Fossil sites, ■. The older, Lake Iroquois site yielded fewer species than the younger, Lake Algonquin site. For example, fossils of the snail *Gyraulus parvus* (on the right) have been found at both sites, but of the clam *Sphaerium striatinum* (on the left) only at the Lake Algonquin site.

in age, and the older, Lake Iroquois site yielded only half as many species as the younger, Lake Algonquin site. Perhaps new immigrants doubled the number of resident species in the interval. Trying to draw conclusions from the absence of certain fossils is not wholly convincing, of course; later studies may show that some species were not really absent after all. Even so, fossils provide firmer evidence of past life than do present-day geographical ranges.

The first proglacial lake to form in the basin of modern Lake Superior appeared about 12k B.P. or soon after. It was a comparatively small lake and is known as Lake Duluth; its position coincided with the western tip of Lake Superior, and it was separated by ice from Lake Algonquin and the lakes farther east (fig. 9.6). Like Lake Agassiz and other western lakes, it drained into the Mississippi, thus, it

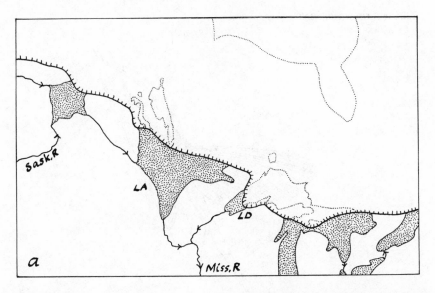

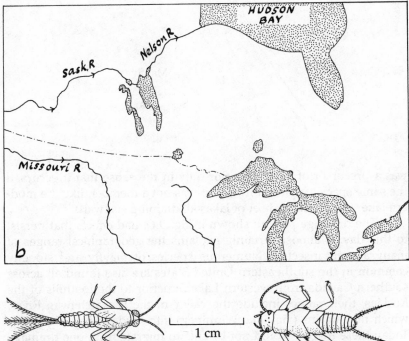

FIGURE 9.6: (*a*) The ice front lakes at about 11k B.P., with the Saskatchewan River (Sask. R) connected to Lake Duluth (LD) and the Mississippi (Miss. R). LA, Lake Agassiz. (*b*) The modern drainage, with the Saskatchewan River flowing into Hudson Bay. Left, mayfly nymph. Right, stonefly nymph.

Mayfly

⊢———————⊣
1 cm

was a precursor of Lake Superior only in the sense that it occupied
the same space (or part of it). But it was not a member, like the modern lake, of a connected set of lakes all draining eastward.

The drainage pattern shown in fig. 9.6a had effects that persist
to this day: it almost certainly explains the geographical ranges of
many aquatic insects.[20] Numerous species of mayfly and stonefly
common in the southeastern United States are also found all across
southern Canada from western Lake Superior to the foothills of the
Rockies; their ranges include the valley of the Saskatchewan River,
which now drains (via Lake Winnipeg) into Hudson Bay (fig. 9.6b).
Today these insects would not be able to migrate from one drainage
basin to the other. Their immature forms are aquatic nymphs, confined to water; the adult insects are feeble flyers and would be most
unlikely to travel from the Mississippi basin to the Saskatchewan basin given the modern drainage pattern, in which the two rivers are
separated by a height of land. When they were linked by a chain of
ice front lakes, migration must have been easy, however.

No doubt many fish species found their way from the Mississippi refugium into Lake Duluth. Northern pike is believed to be one of them. Several rivers flowing into Lake Superior from the south go over barrier falls (falls impassable to fish). In these rivers, northern pike are found only below the falls in stretches of river accessible to fish swimming upstream from the lake. Muskellunge living in the same rivers are found only above the falls. A possible explanation for the separation of these two closely related fish species is that northern pike did not enter the rivers until after the ice had disappeared and the water level had fallen low enough to make the falls impassable; they were therefore confined to lower stretches of the rivers.[21] Muskellunge, which presumably lived both below and above the falls originally, were crowded out of the lower reaches by the invading pike (there is independent evidence showing that this can happen) and now live only above the falls.

The last of the giant proglacial lakes, comparable in size with Lakes McConnell and Agassiz, was Lake Ojibway (fig. 9.7), which was at its largest about 8.5k B.P. It lay east of Lake Agassiz and was probably continuous with it. The north shore of the combined lakes, where they were walled by ice, must have been over 3,000 kilometers long. Lake Ojibway was separated from Lakes Superior and Huron (which had approximately their present form) by a height of land, but like them it drained into the Ottawa River, a tributary of the Saint Lawrence, whereas the two lower lakes (Lakes Erie and Ontario) drained into the Saint Lawrence directly.

The position of Lake Ojibway's southern shore is inferred from the present-day ranges of various aquatic animals, especially the

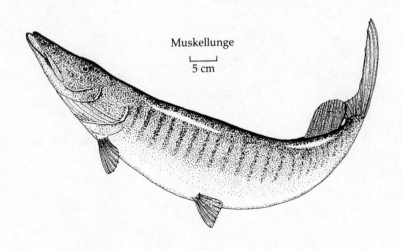

Muskellunge

|___|
5 cm

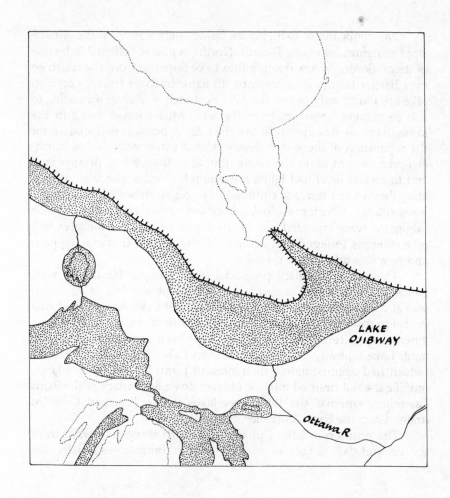

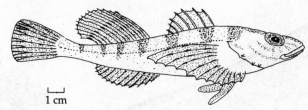
1 cm

FIGURE 9.7: Glacial Lake Ojibway at about 8.5k B.P. Below, four-horned sculpin.

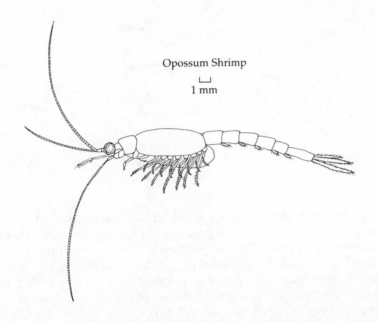

Opossum Shrimp

⌐⌐
1 mm

four-horned sculpin and the opossum shrimp.[22] The opossum shrimp was the sculpin's most abundant food supply. Both species are believed to have migrated into the lake from the south and to have lived in it for as long as it existed. Then, when the lake drained away, the animals were left in the scattering of ponds and small lakes that are all that now remains of Lake Ojibway. They are assumed to have persisted where they survived, but not to have spread, so that a line separating ponds that now contain them from ponds that now lack them represents the ancient shoreline.

Another Lake Ojibway fish was goldeye. It does not occur in the modern Great Lakes, and its main range is to the west of them. But it also occurs in a small disjunct region south of the modern James Bay (the southern "finger" of Hudson Bay) in waters that were once part of Lake Ojibway. The only reasonable explanation for this curious range is that goldeye moved north from the Mississippi basin into Lake Agassiz, and then east into Lake Ojibway, after the link between Lakes Agassiz and Superior had been broken, thus bypassing the Great Lakes.[23]

Lake Ojibway came to a spectacular end about 8.3k B.P.[24] The ice covering Hudson Bay had weakened along a north-south line, and the northwest arm of the lake, shown in fig. 9.7 as just reaching James Bay, lengthened northward. At the same time an arm of the

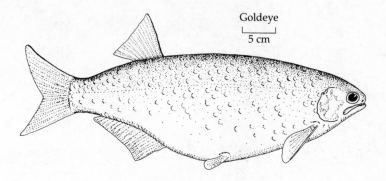

Goldeye

|_____|
5 cm

sea on the north side of the ice sheet lengthened southward, along the line of weakness. The ice separating sea and lake became a steadily narrowing isthmus. The lake surface was 250 meters higher than sea level. Consequently, the isthmus acted as a high dam, and when it finally gave way, Lake Ojibway emptied into the sea with catastrophic suddenness. Thus ended the last of the giant ice front lakes. To the north of it appeared a "new" sea, which we consider in the next chapter.

10

The Rising Sea

Directly or indirectly, the meltwater from the thawing ice sheets reached the sea. When the thaw began, a large proportion of the ice sheet margins were cliffs of ice rising out of the sea; the cliffs calved off ice bergs that melted as they floated away. Later in the thaw, when the ice sheet margins lay mostly inland, meltwater accumulated in proglacial lakes as described in chapter 9. It flowed out from these lakes through the continent's great rivers: the Columbia and Fraser rivers to the Pacific; the Yukon River to the Bering Sea; the Mackenzie River to the Arctic Ocean; the Mississippi River to the Gulf of Mexico; the Saint Lawrence and Hudson rivers to the Atlantic; and the Saskatchewan, Nelson, and Churchill rivers to Hudson Bay, when the bay finally became ice free. The valleys of these impressive rivers owe their existence to glacial meltwater.

The water itself raised the worldwide level of the sea. Ice sheets in Europe and Asia were thawing at the same time, and contributed some meltwater, but they yielded a much smaller volume than did the North American ice sheets. The rising sea submerged big tracts of shoreline land, converting them from coastal plains to continental shelves. Many ice age refugia became, and remain, inundated. Other low-lying regions were temporarily inundated by the sea but are now high and dry.

The tremendous volume of water involved defies the imagination; at one time or another during the melting period about one-fourth of the whole area of Canada was under fresh or salt water.[1] We have already considered the fresh water; this chapter deals with the expanding sea and the ecological consequences of its expansion.

The Sundering of Beringia

One of the most profound changes wrought by the rising sea was the progressive submergence of the land bridge between Alaska and Asia, ending with the creation of Bering Strait, when the land bridge was finally breached (compare figs. 1.2 and 1.3). This happened about 15.5k B.P.[2] As the strait widened, and more and more low-lying land was submerged, the Beringian climate changed. After being dry and intensely cold, it gradually became moister and milder. In a word, more maritime.

The change in climate caused a change in vegetation. The prevailing arctic herb tundra, of grasses and low herbs, was replaced by a more luxuriant shrub tundra dominated by low-growing birches, probably a mixture of dwarf birch and shrub birch. The two kinds of birch intergrade in appearance and size; the lowest dwarf birch

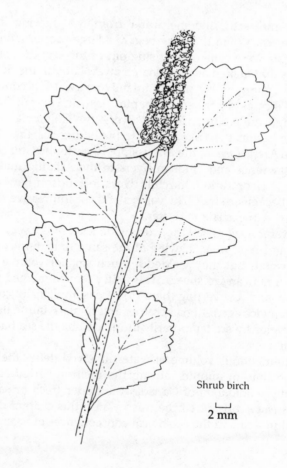

Shrub birch

2 mm

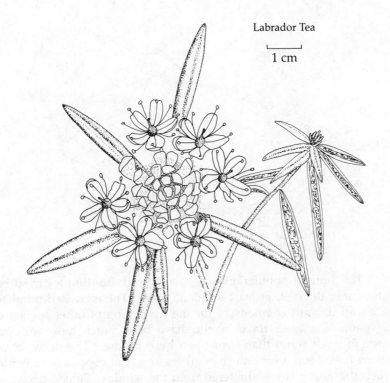

Labrador Tea

1 cm

merely carpets the ground, while the tallest shrub birch can reach a height of two and a half meters. They interbreed so readily that many botanists regard dwarf birch as merely a subspecies of shrub birch. The pollen grains of the two kinds are almost indistinguishable (see Interpreting Pollen Diagrams, chapter 3). With the birch grew other low shrubs: willows, crowberry, blueberries, and Labrador tea.[3]

The switch from herb tundra to shrub tundra happened rapidly; pollen diagrams from widely scattered sites all show the changeover as occurring at about the same time. In addition to large tracts of birch scrub, there were patches of "tussock tundra" dominated by cotton grass and wet places covered by sedge meadows.[4] A variety of flowering herbs and ferns also flourished.

The sudden domination of large areas by shrubs instead of herbs would have greatly altered the appearance of the landscape and must also have affected the well-being of grazing and browsing mammals. Wapiti became relatively more numerous,[5] probably because they are extremely adaptable. They could successfully switch from a diet of grass to a diet of twigs in a way that the habitual grass eaters (mammoths, horses, and bison) could not.

Walrus

|———|
10 cm

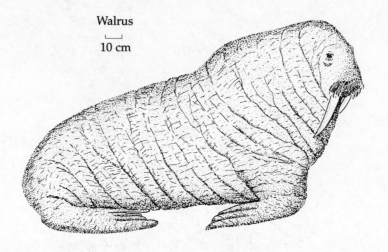

The human population was affected too; hunting for meat became more difficult, at least in inland areas. The increased numbers of wapiti did not compensate for the dwindling of other species of big game. As well, travel would have been much more arduous through birch scrub than over open herb tundra.[6] The dense scrub would have been a hindrance at all seasons, but especially in winter when the heavy snow, sheltered from the wind, remained deep and soft and exhausting to slog through.

But if the environment became less hospitable to humans in the inland parts of Beringia, it probably became more hospitable close to the ocean. Clear skies and bitter cold gave way to foggy drizzle and milder temperatures. Conditions came to resemble those in refugia farther south, along the coast of what is now British Columbia (see The Western Edge of the Ice, chapter 6). The greater warmth allowed marine mammals such as walrus and the now extinct Steller's seacow to migrate north to Alaskan coasts from southern refugia, where they had survived the height of the glaciation.[7] Thus meat supplies for the human population increased on the coast as they became scarcer inland.

The Atlantic Shore

Now let us consider the changing Atlantic shoreline. As described in chapter 6 (see The East Coast Plains and Islands and fig. 6.4), at glacial maximum much of what is now submerged continental shelf was above sea level. From Georges Bank (fig. 10.1) south, there was a widened coastal plain; north of Georges Bank, the exposed land took the form of big, low-lying islands.

Then the ice sheet began to melt, and its eastern margin to retreat westward. The retreat was rapid, at times as much as 300 meters per year.[8] The earth's crust had been depressed by the weight of the overlying ice, so that a large tract of what is now coastal Maine was below what was then sea level, itself more than 100 meters below the present level. The sea flooded in. This so-called *marine transgression* into Maine did not entail the flooding of newly ice-free land. Rather, the land surface was already part of the sea floor, off a coast formed by the ice sheet margin. The present sites of Augusta and Bangor were on the bottom of this expanding bay, which extended some way up the Penobscot valley. The bay is labeled M in fig. 10.1a; it was at its largest soon after 13k B.P.

Once the weight of the ice sheet was removed, the downwarped crust began to rebound. What had been sea floor began to emerge as dry land; a strip corresponding to the modern coastline was the first to emerge, and it formed a chain of islands across the mouth of the

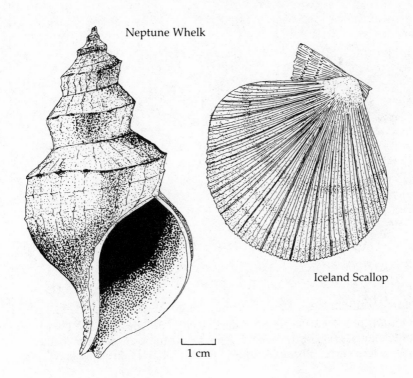

Neptune Whelk

Iceland Scallop

1 cm

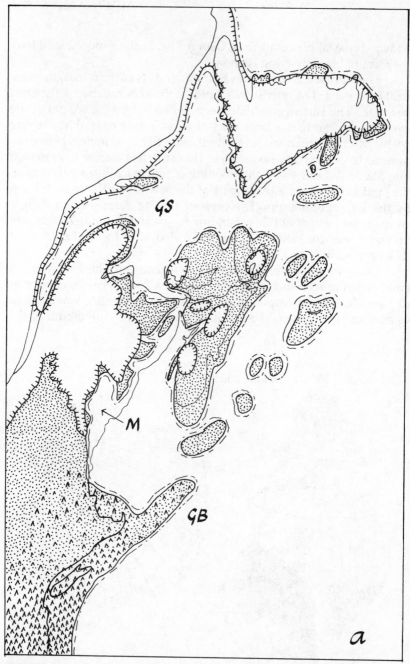

FIGURE 10.1: New England and the east coast of Canada. (*a*) At 13k B.P. (*b*)
at 12k B.P. GS, Goldthwait Sea; M, the marine transgression into Maine;
GB, glacial Georges Bank; BF, Bay of Fundy; NB, New Brunswick; NS, Nova

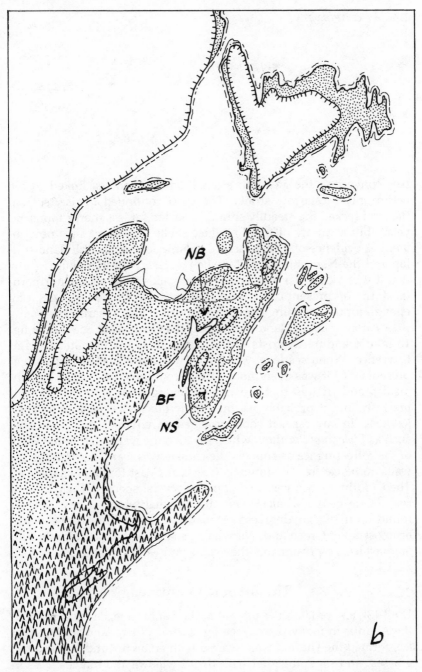

Scotia. The symbols on the land areas represent forest (in the south) grading northward into tundra (stippled area). The large tract of ice is the Laurentide ice sheet.

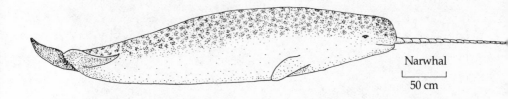

Narwhal

|_____|
50 cm

bay, converting the bay into a small land-locked sea linked to the ocean only by narrow straits. The crust continued to rise, so that the land-locked sea steadily shrank. It lasted for less than a thousand years, but a variety of fossil mollusc shells remain, at sites now inland, as evidence of its existence. Examples are shells of Iceland scallop and the Neptune whelk *Neptunea despecta*. These molluscs show that it was a cold sea with temperatures like those off the northernmost tip of present-day Newfoundland. It did not endure long enough for the warming climate to convert it into a warm sea.

Another almost land-locked sea, the Goldthwait Sea (GS in fig. 10.1*a*) did endure. It exists to this day and is now the Gulf of Saint Lawrence. When sea level was low, it was more land-locked than at present since it was even more cut off from the open ocean, by peninsulas and islands that are now submerged. It was shallow and brackish, and it probably warmed fairly quickly because of its shallowness. In any case, it soon provided an environment for forams such as *Elphidium clavatum*, which invaded the waters left in the wake of the retreating ice as soon as the climate was warm enough for the water to be ice free in summer[9] (see East Coast Refugia, chapter 6). The Goldthwait Sea was also home to several species of whales and seals. One of the whales[10] was the narwhal which is now seldom found south of sixty degrees north latitude, the latitude of the southernmost tip of Greenland. Narwhals usually feed close to ice floes or landfast ice, so presumably they came and went with the seasons.

The Atlantic Coastlands

Until 13k B.P. or later, the margin of the Laurentide ice sheet reached the Atlantic in many places (see fig. 10.1*a*). Where land, rather than ice, formed the shore it was covered with an arctic tundra of sedges, grasses, and sage, together with dwarf willows and birches.[11] As in Beringia, ferns seem to have been more abundant than they are in modern arctic tundra. Fossils of such arctic-alpine plants as dryads and mountain sorrel show that these plants then grew in the lowlands of what is now New England.

Along the coast, the eustatically rising sea could not keep pace with the isostatically rising (rebounding) land. Consequently, the marine transgression into Maine described previously quickly drained away; by about 12k B.P. the coastline of the mainland from Cape Cod to the Bay of Fundy (see fig. 10.1b) was much as it is now. On land, forests slowly advanced northward from the latitude of present-day Long Island, replacing tundra.

Farther seaward, the eustatic rise in sea level exceeded the isostatic fall, but the low-lying offshore islands were not yet submerged; they still existed in shrunken form (compare figs. 10.1 and 6.4). The Atlantic coast of Nova Scotia was still considerably to seaward of its present position, and glacial Georges Bank was still high and dry; mastodons and mammoths roamed over its tundra and browsed in its sparse spruce woods (see chapter 6). The ice-free offshore islands and the expanded continental shelf also served as refugia for numerous animals and plants that survive to the present day; they migrated back onto the mainland when it became available.

Thus the modern vegetation of peninsular Nova Scotia has descended in part from ancestors that lived "out to seaward," on land now submerged, and in part from ancestors that lived south of the ice sheet and gradually migrated northeastward, as conditions improved, along the coastal plain and around the head of the Bay of Fundy.[12] Many plant species probably reached the peninsula from both directions, but some must have come only from the seaward side. These are species now found as disjuncts, whose range in Nova Scotia is limited almost entirely to the western end of the peninsula. They have evidently been unable to spread eastward, presumably because of competition from plants that got there first. Two examples are catbrier and skunk cabbage (fig. 10.2).

Many species of trees invaded from both directions. The populations that followed the mainland route have left evidence of their time of arrival in the form of pollen preserved in lake sediments. They advanced northeastward through New England from their refugia in the south. First to come were poplars, followed by spruces, balsam fir, and paper birch, trees that can endure extreme cold. Then came less hardy trees: oak, maple, eastern white pine, and eastern hemlock established themselves, in that order.[13]

The order was undoubtedly governed to some extent by their dissimilar ecological requirements; as noted in chapter 4, eastern hemlock would be expected to lag behind eastern white pine because the deeper, moister soil it needs would have taken time to form. But an equally important factor in determining when each immigrant species managed to establish itself was ecological inertia (see chapter

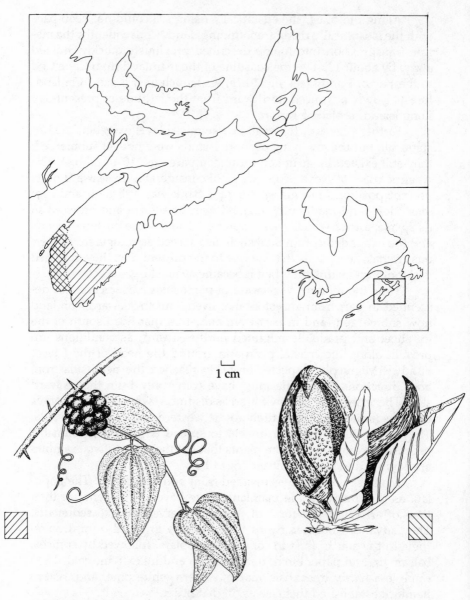

1 cm

FIGURE 10.2: The present ranges of two plant species that presumaɒly invaded Nova Scotia from offshore island refugia and that have been unable to spread. Left, catbrier. Right, skunk cabbage.

4). The evidence comes from the presence of layers of charcoal fragments, as well as pollen, in sediment cores. Cores from a lake in
southwestern Nova Scotia[14] show that every time one species of tree
replaced another as the dominant species in the surrounding forest,
there had been a fire (as shown by charcoal fragments) at the time of
the changeover. The "new" species was probably present in the
neighborhood in small numbers long before the changeover; environmental conditions may even have come to favor it more than the established species. But it could not become dominant until fire destroyed the existing forest and made space for it.

The Champlain Sea

While the events just described were happening along what is now
the east coast, the coastline itself was changing dramatically. We have
already considered one, rather minor, marine transgression, into
coastal Maine. A vastly greater one happened when the ice blocking
the Saint Lawrence valley disappeared and a water route was suddenly opened up into the interior of the continent. Large inland areas
were below sea level because of crustal downwarping; the ocean
flooded in and the Champlain Sea came into existence. It filled all the
Saint Lawrence valley, much of the Ottawa River valley, and the basin
of modern Lake Champlain. The present sites of Ottawa, Montreal,
and Quebec City were all submerged (fig. 10.3).

The Champlain Sea lasted for 2,000 years, from about 12k to 10k
B.P.[15] while the crust gradually rebounded. Early in its existence it
was much less saline than the open ocean and exceedingly cold. Its
north shore consisted of melting ice cliffs, and over its surface drifted
expanses of pack ice, interspersed with towering bergs. A variety of
marine mammals lived in its waters.[16] There were five species of
whales: bowhead, humpback, finback, beluga, and harbor porpoise.
And three species of seal: ringed, harp, and bearded. All but the
humpback and finback whales and the harbor porpoise are animals
of high latitudes, hardly ever traveling farther south than the Gulf of
Saint Lawrence at the present day.

These large mammals require quantities of animal food. There
must have been rich supplies of zooplankton, the food of humpback,
finback, and bowhead whales. These are baleen whales, which use
filters of whalebone (baleen) to strain plankton from large volumes of
sea water. The plankton probably contained plentiful krill, the collec-
tive term for species of euphausids, small, red, shrimplike crustaceans living in surface waters. The other two whale species, belugas
and harbor porpoises, are toothed whales, which feed on a variety of

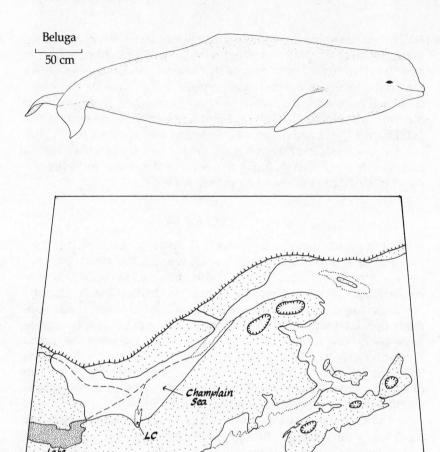

FIGURE 10.3: The Champlain Sea, as it was at about 10.5k B.P. The broken lines show modern rivers: the Ottawa River, the Saint Lawrence River, and the Richelieu River flowing from Lake Champlain (LC) into the Saint Lawrence. Below, harp seals on an ice floe, with humpback whales in the distance.

foods such as fishes, squids, octopuses, and crustaceans. Harp and ringed seals have much the same diet as the toothed whales, and eat krill as well; bearded seals are bottom feeders that eat whelks, clams, crabs, octopuses, flounders, cod, and sculpins. The Champlain Sea undoubtedly developed a diverse marine ecosystem. Fossils show that among the fish that served as fodder for the seals and toothed whales were a species of codfish, tomcod,[17] and three-spined stickleback. Remains of soft seaweeds including kelp have also been found.

Conditions in the Champlain Sea varied markedly during its short lifetime because water reached it from different sources at different times. Sometimes it was cooled and diluted by frigid meltwater from the ice sheet; sometimes it was diluted by fresh water from comparatively warm lakes; and sometimes it was warmed and made more salty by influxes from the Gulf of Saint Lawrence. And all the time it was becoming steadily shallower as its floor rebounded.

The changing abundances of different species of clams, forams, and ostracods that lived on the floor of the sea, and are now embedded in the sediments, provide a record of these changes.[18] The environment appears to have changed abruptly soon after 11k B.P. The dominant clam before this time was the arctic saxicave, a species found in frigid waters. The saxicaves suddenly disappeared and were replaced by soft-shell clams, which require warmer water than do saxicaves if they are to reproduce successfully. The ostracod species living with these clams show how the salinity changed from highly saline during the saxicave period to brackish during the soft-shell

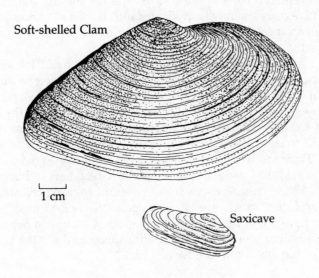

Soft-shelled Clam

1 cm

Saxicave

chipmunk

└─────┘
2 cm

clam period. There must have been a sudden influx of fresh water, perhaps from glacial Lakes Iroquois and Algonquin (see fig. 9.5).

Terrestrial life on the lands surrounding the Champlain Sea presumably did not change so abruptly. The sea was probably ringed by tundra when the ice margin first drew back from its shores. Tundra was then replaced by boreal forest. As the sea shrank, the forest advanced to occupy the newly available land. Fossils of various animals that lived in these forests have been found, among them snowshoe hare, marten, and eastern chipmunk.[19]

The shores of the sea provided an environment for shoreline plants and animals and evidently served as a link between the ocean shores of the Atlantic and the freshwater shores of the Great Lakes. For many shore species, it makes no difference whether the nearby water is fresh or salt; they are adapted simply to a habitat of loose sand. Three such plants are sea rocket, beach grass, and seaside spurge. Their modern ranges are disjunct; they are found along the Atlantic coast and around the southern Great Lakes, but not between.[20] Although inferences from modern geographical ranges are never absolutely conclusive, there seems little reason to doubt that they did have a continuous range when the Champlain Sea existed, with shores that were continuous with those of the Atlantic to the east and Lake Ontario to the west.

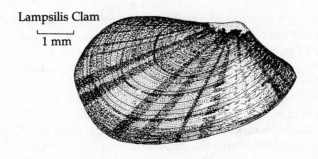

Lampsilis Clam

1 mm

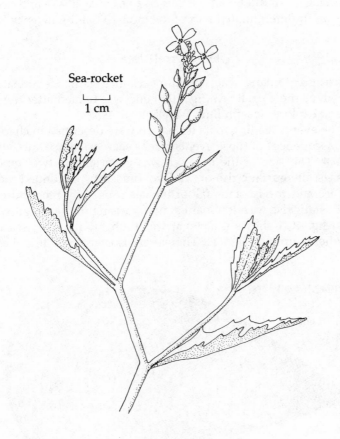

Sea-rocket

1 cm

As the crust rebounded, the Champlain Sea became smaller and smaller. Finally its eastern outlet was above the level of the sea outside (the Gulf of Saint Lawrence), and what had been a salt sea was converted to two freshwater lakes. They were fed by Lake Ojibway (see chapter 9) and the precursors of today's Great Lakes. The smaller of the two still survives; it is modern Lake Champlain (see fig. 10.3). The larger, which no longer exists as a lake, is known as Lampsilis Lake, after *Lampsilis radiata*, a species of freshwater clam that lived on its shores[21]; these clams are still common throughout the Great Lakes region.

Lampsilis Lake did not last long. As the crust continued to rise, the lake became steadily narrower until it was no wider than the river that flowed into its southwestern end and out of its northeastern end. At this stage, around 8k B.P., it ceased to exist as a lake and became merely an undifferentiated part of the modern Saint Lawrence River.

The Tyrrell Sea

One last, great inland sea came into existence as the ice retreated. It is known as the Tyrrell Sea (fig. 10.4) and was named after the geologist and explorer Joseph Burr Tyrrell (1858–1957).

The events leading to its formation were described in chapter 9. The consequences of those events can be seen by comparing figs. 9.7 and 10.4. The size of the new sea was governed by two opposing processes: the eustatic rise of sea level caused by continued melting of the ice and the isostatic fall of the sea caused by rebound of the land. Eustatic rise was dominant at first, causing the sea to grow. The map in fig. 10.4 shows it at what was probably its greatest extent, somewhere around 7k B.P. Thereafter, isostatic rebound became

Blue Mussell

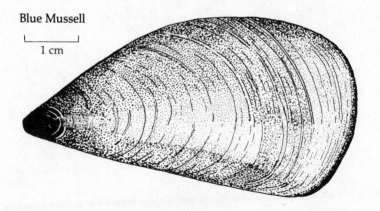

|_____|
      1 cm

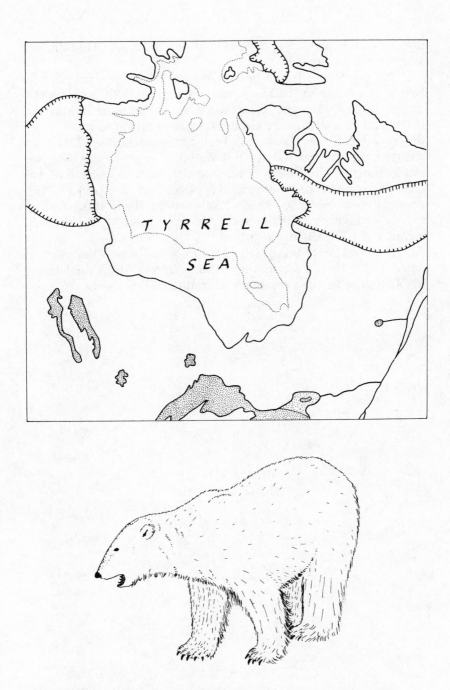

FIGURE 10.4: The Tyrrell Sea, circa 7k B.P. The dotted line is the modern
Hudson Bay coastline. Below: Polar bear.

dominant; the Tyrrell Sea decreased in size to become Hudson Bay, which is decreasing still.

No doubt the Tyrrell Sea closely resembled modern Hudson Bay. It was home to the same marine mammals, fish, and invertebrates. Its low, flat shores supported a patchwork of tundra and marsh and provided a land habitat for polar bears in summer, when the rotten sea ice made seal hunting impracticable. Little has changed, except that the sea has shrunk. Tier upon tier of raised beaches around the shores of Hudson Bay show how sea level has fallen. Blue mussel shells are found at places now far inland and high above present mean sea level: 2,300-year-old shells at a site now twenty-six kilometers inland and 1,400-year-old shells at a site now eleven and a half kilometers inland.[22]

Sea level continues to fall as the floor of Hudson Bay rises; as noted in chapter 1, it is a toss-up which will come first: a total drying up of Hudson Bay or a return of the continental ice sheets.

PART FOUR

THE PLEISTOCENE/ HOLOCENE TRANSITION

11

The End of an Epoch

The End of the Pleistocene

Chapters 9 and 10 gave an outline history of the ice front lakes and inland seas that came into existence, one after another, in the wake of the melting ice sheets; the account ended at 7k B.P. when the Tyrrell Sea reached its greatest extent. We now backtrack 3,000 years to the year 10k B.P. The year is of no special significance in the history of the continually changing pattern of land and water, but it is of great significance in other contexts.

First, the year 10k B.P. marks the "official" end of the Pleistocene epoch (see chapter 1). That is to say, it has been *defined*, by professional geologists (specifically, by a subcommission of the International Quaternary Union, known as INQUA), as the date of the Pleistocene/Holocene transition. The year 10k B.P. is therefore epoch making by definition.

Second, the year 10k B.P., or at any rate a fairly short interval centered on that year, *may* mark the end of an epoch in the ordinary sense. As noted in chapter 9, the terrestrial biosphere underwent some rapid and important changes at approximately 10k B.P., which are believed to signal a sudden, major climatic change. Three of these abrupt occurrences, as they could be called, deserve particular mention.

The first of them[1] has to do with the amount of carbon dioxide in the atmosphere, which has fluctuated widely over the millennia. The atmospheric concentration of carbon dioxide in times past can be discovered by taking cores of ice from existing ice sheets. The ice consists of datable annual layers, each containing small air bubbles representative of the contemporary atmosphere; the carbon dioxide content of the air in the bubbles can be measured. Ice cores collected at Camp Century in northern Greenland and at Byrd Station in Antarctica have been analyzed in this way, and it turns out that carbon dioxide concentration increased steadily, from a value of about 210 parts per million at roughly 17k B.P. to about 280 parts per million at roughly 10k B.P. (note the unavoidable imprecision). The value remained close to 280 parts per million thereafter, until the human population and industrial explosion of our own time caused the value to start climbing again. Carbon dioxide concentration was therefore increasing at the same time (from 17k to 10k B.P.) as the climate at high northern latitudes was warming. The increase could not have been the sole cause of the warming, however; it would probably have caused a worldwide warming of no more than 1.4° Celsius, whereas the average temperature of the atmosphere is believed to have risen at least 2.3° Celsius during the last 7,000 years of the Pleistocene.[2] It is even possible that rising temperatures were a cause, rather than an effect, of increasing carbon dioxide. Be that as it may, the fact remains (provided the observations are reliable) that, at about 10k B.P., there was an abrupt occurrence, namely, an end to a long-continued rise in carbon dioxide concentration.

The second abrupt occurrence, also of worldwide import, seems to have happened slightly earlier, at 11k B.P. It is marked by a sudden change in the kinds of foraminifera found in sediments on the floor of the ocean. Many species of forams have spirally coiled shells, and a single species can exist in two varieties, some with left-coiled spirals, the rest with right-coiled spirals. In some of these species, the two varieties are adapted to different water temperatures; for example, the left-coiled variety may live only in cold arctic seas, and the right-coiled variety only in warm temperate seas. The forams in some carbon-dated sediment cores from the ocean floor show an abrupt change in coiling direction, indicating an abrupt rise in water temperature at the 11k B.P. level.[3] This is strong evidence for a sudden climatic warming.

The third abrupt occurrence—abrupt, at least, in the geological sense—was the extinction of a great many species of North American mammals.[4] Over fifty species went extinct during the whole course of the Wisconsin glaciation, but most of the extinctions happened in one short burst at the end, in the interval 12k to 10k B.P. (assuming that each species disappeared soon after the date of its most recent known fossil). According to some researchers,[5] the wave of extinctions ended at 10k B.P., and no mammals have gone extinct on this continent since. This certainly makes 10k B.P. the end of an epoch in the colloquial sense, the epoch being that of the great ice age mammals.

The great wave of extinctions is one of the most important and

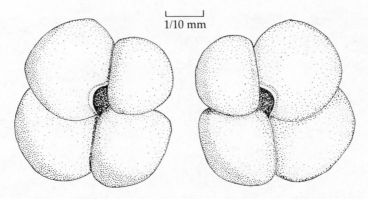

1/10 mm

Left- and Right-coiled Forams (Globigerina pachyderma)

mysterious ecological events in the last 100,000 years and is discussed in detail in chapter 12. The present chapter considers the setting in which these extinctions took place. It provides a snapshot of the environment as it was around 10k B.P., at what could be called "the turn of the epoch."

The Changing Forest

At about the same time as the abrupt occurrences just mentioned, there was an equally abrupt change in the composition of the forests, woodlands, and parklands south of the ice sheets. Spruce, mostly white spruce, after being by far the most abundant species, suddenly became rare. In its place a much more diverse forest developed, made up of three species of pines—jack pine, red pine, and eastern white pine—together with balsam firs, and a variety of hardwoods, chiefly paper birches, elms, and oaks.

Pollen diagrams showing the abrupt change come from numerous sites: Minnesota, Wisconsin, Iowa, Ohio, Michigan, Pennsylvania, Connecticut, Massachusetts, New Brunswick, Nova Scotia, southern Ontario, and Quebec (fig. 11.1). The sudden spruce decline was apparently synchronous over a huge region, stretching from Minnesota to Nova Scotia.

The rapidity of the spruce decline in the Midwest was remarkable. At some sites it happened in less than a human lifetime,[7] and small bands of Paleo-Indians, probably caribou hunters, may have been there to witness the change. The disappearance of spruce left the land available for replacements—pine, fir, and northern hardwoods—that had been steadily expanding their ranges northward

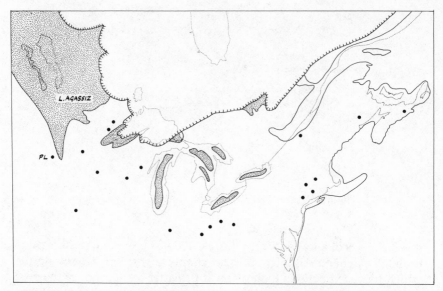

FIGURE 11.1: The land from Lake Agassiz to the Atlantic at 10k B.P. ●, sites where pollen diagrams show a sudden replacement of spruce by other tree species (pines, fir, birch, oak) in a short period centered on 10k B.P.[6] The sites have remained forested to the present day, except for the westernmost, Pickerel Lake (PL), where prairie grassland soon replaced forest.

over thousands of years. Each species advanced at its own rate. Jack pine, the fastest migrant, took two or three thousand years to spread into the Midwest from its refugium in the Appalachian highlands and Atlantic coastal plain; as jack pines are "pioneering" trees, able to flourish on dry, sandy soils, it seems likely that they migrated west along old shorelines left by ice front lakes. Pollen diagrams show that as spruce woods gave way to mixed forest, the herbs, grasses, and sedges dwindled sharply: open, grassy woodland with well-spaced trees was being replaced by closed, shady forest.[8]

To understand how the forests developed, palynologists consider both pollen percentage diagrams and pollen influx diagrams; the latter show the numbers of pollen grains of each species deposited per square centimeter per year (see chapter 2). Such studies have shown the course of events in New England.[9] There was a marked contrast between northern and southern parts of the region. Following the disappearance of the ice, spruces were the first trees to appear, and increasing quantities of spruce pollen show how in southern New England, spruce forests burgeoned as the climate warmed

Eastern White Pine

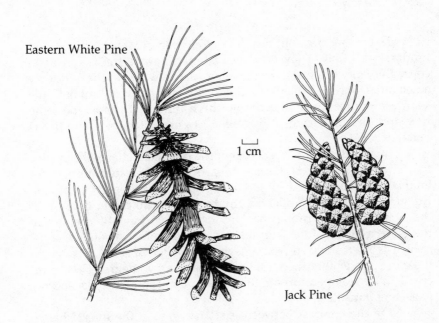

1 cm

Jack Pine

Lodgepole Pine

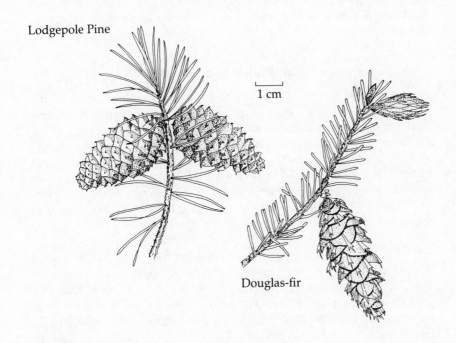

1 cm

Douglas-fir

from 12k B.P. onward. But this did not happen in Maine, where open tundra, with only a few, widely scattered spruce trees, spread to cover the land when the encroaching sea drew back. The suddenly accelerated warming of 10k B.P. led to the disappearance of both the southern spruce forest and the northern tundra at almost the same time and their replacement by forests of pine, birch, oak, and (later) hemlock.

The forests on the west coast of the continent changed abruptly, too. The first postglacial forests in parts of southwestern British Columbia were dominated by lodgepole pine. But then, at 10.5k B.P. Douglas-fir suddenly appeared in abundance, no doubt because of the same abrupt warming that produced such dramatic changes in the eastern forests. Inferring climatic change from vegetation change (which is itself inferred from changes in pollen spectra), has led to the conclusion[10] that there must have been a two-degree Celsius increase in July temperatures in the 500 years preceding the Pleistocene/ Holocene transition.

Note the contrast between east and west in the forest change that took place. Pines, especially jack pine, were beneficiaries of the climatic change in the east, where they took over from spruce. In the west, lodgepole pine (which is very closely related to jack pine), was the loser. It was to a large extent displaced by Douglas-fir when the climate warmed.

The Prairie Grasslands

Between the eastern and western forests, in the rain shadow of the western mountains, lie the prairie grasslands. These lands have gone through more stages than any other since the disappearance of the ice, having been tundra covered, and then forested, before grasses took over. At least ninety percent of the present-day prairies was under water—submerged by an ice front lake—or under stagnant ice for at least a few decades following the disappearance of active ice.[11]

The land that first became available for terrestrial vegetation was therefore certainly wet. It was either a newly drained lake bed or a tract of stagnant ice terrain riddled with pools of meltwater (see chapter 8). Precipitation must also have been greater then than it is now; the proximity of the ice front lakes compensated to some degree for the midcontinental site. The water was ample for trees, and tree growth became possible as soon as the layer of soil that thawed in summer had become deep enough for their roots. Muskeg was plentiful in the first forests[12] (see chapter 4). There was also muskeg in the region that had never been ice covered, south of the southern-most limit reached by the Laurentide ice sheet.

The drying out of the muskegs and their replacement by grasses was a gradual time-transgressive process that affected all the Great Plains, from far south of the ice margin to as far north as the northern limit of the prairies. In the context of the prairies, the Pleistocene/Holocene transition at 10k B.P. is an arbitrary date. All that can be said of it is that it probably comes at about the midpoint of the prairie advance, which lasted from about 12k B.P. to about 8k B.P. Note the *abouts*. As remarked in chapter 5, lakes with pollen-bearing sediments providing a long, continuous record are rare in the arid west, in part because of the aridity; the history of the changing vegetation has to be based on rather scanty records.

A sediment core from Pickerel Lake in the northeastern corner of South Dakota has given one picture of the change from forest to grassland[13] at a place where the change came slightly earlier than the formal Pleistocene/Holocene transition. Figures 11.1 and 11.2 show the location of Pickerel Lake. It is at the easternmost limit of present-day prairie, and when it first came into existence on top of stagnant ice, at about 10.6k B.P., the southern tip of Lake Agassiz was not far away. Coniferous forest surrounded it, much of it muskeg dominated by black spruce and tamarack. Only a century passed before the vegetation changed abruptly, becoming much richer. The ground dried, perhaps because of a local drainage change, perhaps because of a drier climate. Balsam fir and paper birch invaded, to be followed by

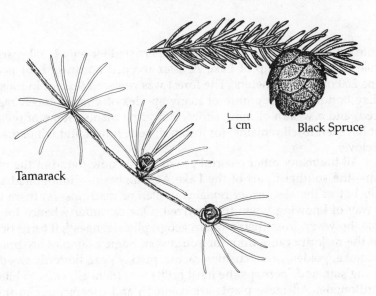

1 cm   Black Spruce

Tamarack

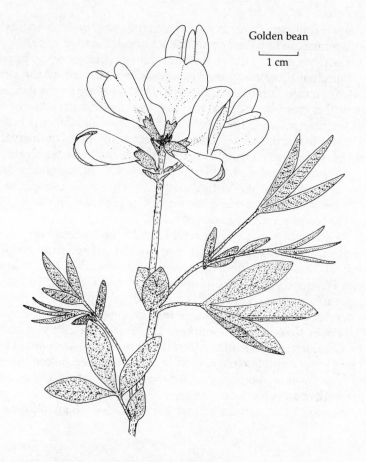

Golden bean

└─── 1 cm ───┘

white elm and burr oak. But this forest persisted for only 1,000 years, while the southern tip of Lake Agassiz receded northward for perhaps 200 or 300 kilometers. The forest was replaced (except in moist valley bottoms) by a prairie of many species of grasses, sage, ragweed, and a wealth of more showy flowering plants. Some of these last have left fossil remains, for instance, cone flower and white prairie clover.

All the many other colorful prairie plants now found in the region—the southern part of the Lake Agassiz basin—immigrated as well, but in the absence of remains (pollen or macrofossils) there is no way of knowing when they arrived. One can infer whence they came, however, from their modern geographical ranges[14]; it turns out that the majority came from the southwest. Some examples are prairie smoke, golden-bean, cushion cactus, prickly pear, butterfly weed, blazing star, and, perhaps the most brilliant of them all, smooth blue beardtongue. All these plants are common and conspicuous in the prairies today.

The farther west we go, the drier the country becomes, and the fewer the lakes with pollen deposits yielding a continuous history of the vegetation. Some other sites are known[15] where, as at Pickerel Lake, deciduous forest replaced spruce forest only to be replaced itself, fairly soon, by prairie. And there are still other sites[16] where, in contrast, prairie succeeded spruce forest without any intervening deciduous forest period (fig. 11.2).

The retreat of the Laurentide ice sheet and its attendant ice front lakes was roughly toward the east-northeast. Therefore, postglacial warming and drying began first in the westernmost parts of the

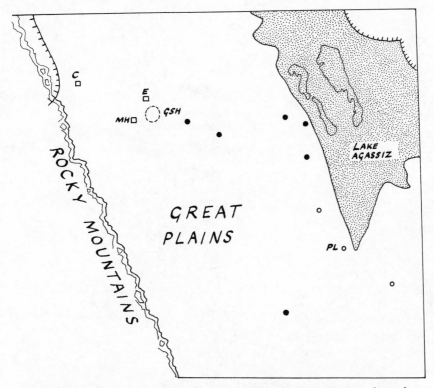

FIGURE 11.2: The circles mark sites where prairie replaced spruce forest between 12.5k and 9.5k B.P.: ○, sites where spruce forest was followed by deciduous forest and later by prairie; ●, sites where there was no deciduous forest interval. PL, Pickerel Lake. □, sites where fossils of large mammals have been collected: E, Empress, C, Cochrane, MH, Medicine Hat, all in Alberta; GSH, Great Sand Hills (the only once glaciated site where Ord's kangaroo rat is now found). Ice and lakes appear as they were (speculatively) at about 10k B.P.

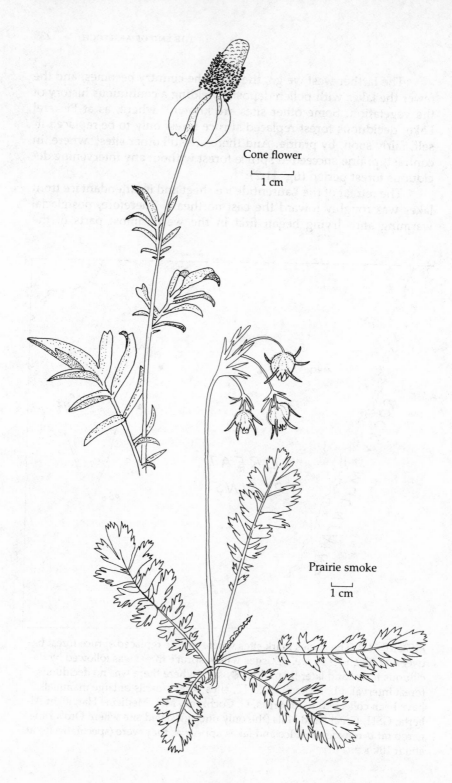

Cone flower

1 cm

Prairie smoke

1 cm

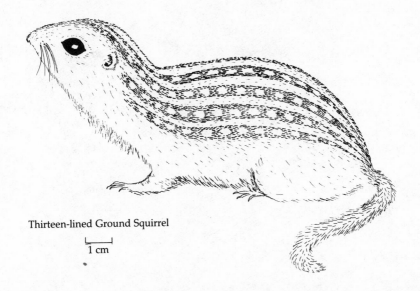

Thirteen-lined Ground Squirrel

∟——⌐
1 cm

plains, immediately east of the Rockies, and progressed eastward in the wake of the receding ice. The farther west we go, the older the prairies, in the sense of having been longest established.

In parts of modern Alberta, fossils of large mammals give indirect evidence of what the vegetation, and hence the scenery, must have been like. Three sites are particularly interesting[17]: Cochrane, Empress, and Medicine Hat (fig. 11.2). Fossils of Mexican wild ass and bison, probably the extinct western bison (see chapter 7), have been found at all three sites and a species of camel, probably western camel, at Empress and Medicine Hat. These animals seem to imply grassland, but there were others more suggestive of tundra, such as caribou at Empress and Cochrane and woolly mammoth at Empress and Medicine Hat. Probably these big herbivores shared an open environment with plenty of grass. The Cochrane and Medicine Hat fossils date from about 11k B.P., and as grasslands must have become established before grazers moved in from their refugium in the unglaciated southern plains, the land may have been under prairie as early as 12k B.P.[18] The date of the Empress fossils is uncertain.

With extant mammals, as with plants, their modern geographical ranges show where they probably came from, though not, of course, when. The early grasslands, where the big grazers roamed, were undoubtedly home to an assortment of small rodents and hares, and also their predators. Judging from their present ranges,[19] some came from the nearest grasslands, those immediately south of the belt of coniferous forest that bordered the ice; examples are white-

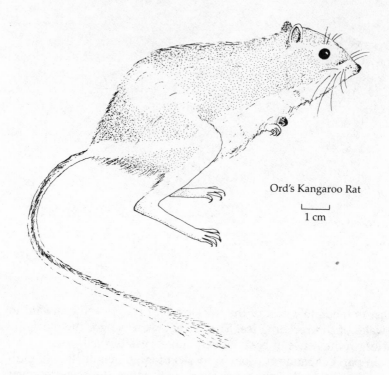

Ord's Kangaroo Rat

|___|
1 cm

tailed jackrabbit, swift fox, and thirteen-lined ground squirrel. (Figure 12.2 shows the present range of the thirteen-lined ground squirrel.) Others came from the more distant southwestern desert, for instance, the western harvest mouse and Ord's kangaroo rat. The kangaroo rat has invaded only a tiny part of the land that was once under ice, namely the truly desertlike Great Sand Hills of southwestern Saskatchewan and southeastern Alberta[20] (see fig. 11.2).

There is no way of knowing when the ancestors of these animals migrated north, but for some others of the small prairie dwellers it is possible to say, within limits, when they came; we return to the topic in chapter 13.

Transition in the West: The Interior

The land under the Cordilleran ice sheet was, and still is, immensely more varied, in the topographical sense, than that under the Laurentide sheet. In place of immense, almost featureless plains, we have a number of rugged mountain ranges, separated by deep, north-south valleys and, at the western edge of the continent, a steep shoreline, cut through by the valleys of several big rivers flowing from the interior mountains into the Pacific Ocean. We now consider some of the changes that came with the Pleistocene/Holocene transition—that is, the transition from glacial to postglacial conditions—while the Cordilleran ice sheet was disappearing. Dates are less certain than places, so rather than attempting a chronological account it is easier to survey the scene from east to west, beginning with the ice-free corridor and ending with the coast.

The ice-free corridor, strictly speaking, belongs neither to the east nor the west; it is the dividing line between east and west, both in the glaciological sense and the biological sense. As noted in chapter 7, the date when a gap first appeared between Laurentide and Cordilleran ice is not known (assuming that they coalesced at some time, which is probable but not certain). Still less is it known when conditions first became hospitable enough in the corridor to allow migration of different kinds of ecosystems into it and through it. Modern vegetation east of the Rocky Mountains gives no hint that there ever was a corridor; the boundary between prairie and forest cuts across it at right angles. Evidently prairie invaded from the south to fill the southern part of the corridor, and boreal forest invaded from the east, along the retreating edge of the Laurentide ice sheet, to fill the northern part.

About one place in the ice-free corridor it is possible to be more specific. The place is Charlie Lake Cave near Fort Saint John, British Columbia (shown on the map in fig. 11.4), which has yielded fossils showing how ecological conditions, and communities, changed at a well-defined date.[21]

Several bird fossils are present, which is rather unusual. In a lower layer, dating from 11k to 10k B.P., are fossils of ground squirrel (Columbian or Richardson's), marmot, a large hare (arctic hare or jackrabbit), and the remains of cliff swallows' nests. Also the bones of a large bison that had been butchered at human hands, and some stone artifacts, including a Clovis-type fluted spear point.[22] In a higher layer, dating from 10k to 9k B.P., are fossils of voles, muskrat, a weasel, ground squirrels, bison, and a variety of birds: western grebe, horned grebe, ruddy duck, sora, coot, and short-eared owl. The changing animal community is taken to imply a change in the vegetation from open country to forest.

Next we turn to the mountains themselves. They gradually became available for life as the Cordilleran ice wasted away. Exactly how the ice wasted is hard to envisage because crucial glaciological questions have yet to be answered. Did the Cordilleran ice consist of an enormous dome (or domes) from which the ice flowed outward in all directions and in which the mountains were deeply buried? Or was

American Coot

Horned Grebe

it an array of comparatively small mountain ice caps linked by enlarged valley glaciers and with plenty of ice-free patches? The speed
at which life returned to the mountains must certainly depend on the
answer to these questions.

In any case, life—both plant and animal—invaded from south
and north. The present alpine flora of the western mountains contains a mixture of plant species, some presumably from refugia in the
mountains south of the Cordilleran ice, others from the Beringian
tundra (only "presumably," bearing in mind that inferences based on
modern geographical distributions can never be as certain as those
based on fossils). As would be expected, the proportions in the mixture vary geographically: invaders from the south become increasingly uncommon the farther north you go. Some examples of "southern" plants are[23] blue camas, bear grass, mountain-box, sticky
currant, Utah honeysuckle, and mountain bog gentian.

Among the immigrants from Beringia[24] are mountain marsh
marigold, mountain monkshood, partridge-foot, mountain harebell,
and golden saxifrage, which besides migrating southward in the
western mountains has also spread eastward across the northern Canadian tundra (see chapter 5).

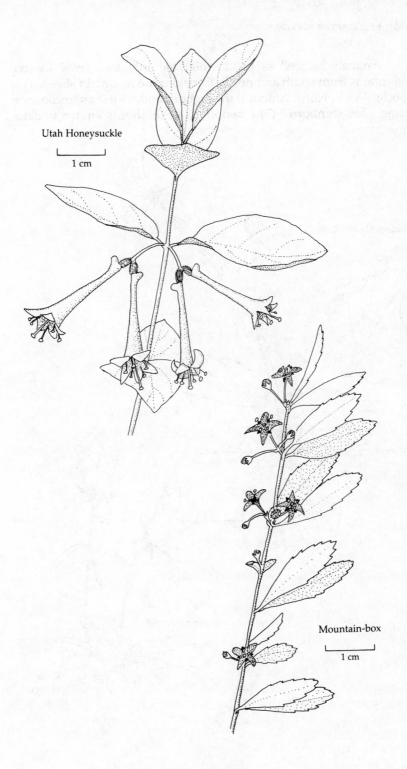

Utah Honeysuckle

1 cm

Mountain-box

1 cm

Animals as well as plants migrated into the newly ice-free mountains from south and north. Consider the mountain sheep. Two species live in North America; the bighorn and, to use an uncommon name, the "thinhorn." (The two kinds of thinhorns known to natu-

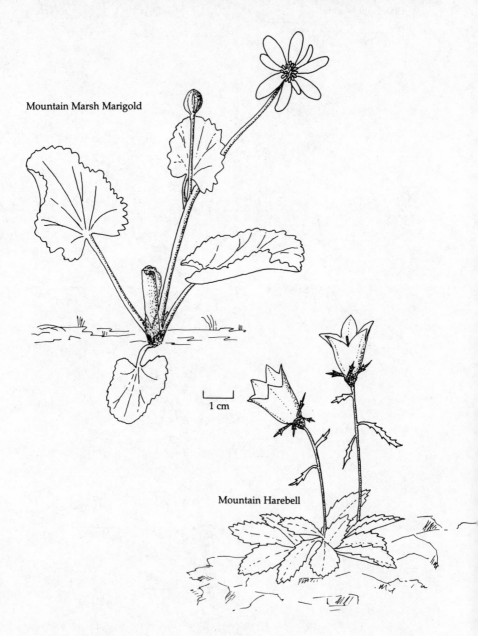

Mountain Marsh Marigold

1 cm

Mountain Harebell

ralists and hunters are treated by zoologists as subspecies; they are the all-white Dall's sheep and the gray-backed Stone's sheep). Figure 11.3 shows the modern ranges of the two species, which certainly suggests that the glacial refugia were Beringia for the thinhorns and

FIGURE 11.3: The modern ranges of bighorn sheep (to the south) and thinhorn sheep (to the north). ×, the site where the 9,300-year-old fossil bighorn skull was found.

the mountains south of the ice for the bighorns. The two species seem not to have met and mingled, even though a fossil bighorn skull dating from 9.3k B.P. has been found to the north of the modern range of the species at the point marked in figure 11.3.[25] The discoverers speculate that the sheep whose skull they found, which belonged to a mature ram, had become trapped between blocks of stagnant ice or ice-cored moraine.

Still on the subject of bighorns, it is interesting to note that bighorn fossils were found at Cochrane, Alberta, along with the fossils of Mexican wild ass, bison, and caribou mentioned previously (and see fig. 11.2); the fossils date from about 11k B.P. Cochrane is out in the plains, about thirty kilometers east of the Rocky Mountains, making the habitat a surprising one for bighorns. Two explanations are possible:[26] either the bighorns of 11k B.P. *did* live in the lowlands of the ice-free corridor while their more usual (to our eyes) environment was under ice; or the fossils were carried from the mountains by a river and redeposited at the site where they were found.

Transition in the West: The Coast

The return of forests to the valleys and steep mountain slopes on the west side of the continent differed markedly from their return to the lowlands and gently undulating slopes of the east. In the mountains, a wide array of different climates can be encompassed in a small area; on the coast, too, the climate was probably mild, even quite close to the ice margin, because of the proximity of the ocean. Probably most of the tree species of the modern West Coast rain forest survived the glaciation near the ice in the valleys of the Coast Range and the Cascades.[27] If they did, they had not far to go to colonize the land newly exposed when the ice melted; all were able to invade promptly, with none of the long delays so characteristic of tree migrations in the eastern half of the continent (see chapter 4). The vanguard of the advancing West Coast forest may have begun growth on stagnant ice; evidence[28] shows that stagnant ice remained in Puget Sound for thousands of years after the active ice margin had receded northward.

As noted previously, the western forests, like the eastern, changed abruptly around 10k B.P. Pollen cores from southwestern British Columbia[29] and adjacent Washington show that lodgepole pine declined suddenly, to be replaced by Douglas-fir; with the Douglas-fir pollen is pollen of red alder and spores of bracken. The findings imply that by 10k B.P., the climate had not only become warmer than before but also considerably drier; forest fires must have been frequent. Areas laid bare by fire provided an ideal habitat for bracken; short-lived red alder then colonized the bracken-filled clearings, improving the soil as it grew (it is a nitrogen fixer; see Conditions in the Newly Deglaciated Land, chapter 4). Then Douglas-fir succeeded the alder. Frequent fires kept the cycle repeating.

While vegetation was changing on the land, a variety of fish species were migrating into the newly ice-free rivers. Cordilleran ice had covered four important drainage basins, those of the Fraser, Skeena, Nass, and Stikine rivers in British Columbia (see fig. 11.4). All flow westward into the Pacific, and all were easily accessible to fish able to tolerate salt water as well as fresh, which could migrate coastwise from their refuges south of the ice. The nearest unglaciated drainage basin, that of the Columbia River, must have been the refuge for many species. For freshwater species unable to tolerate salt

Bracken

⌞⌟
1 cm

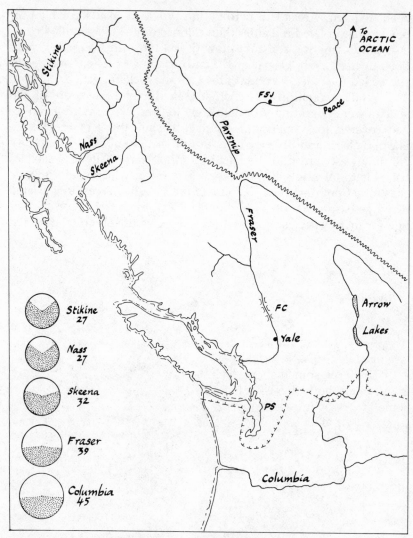

FIGURE 11.4: The pie diagrams on the left show the proportions of salt-tolerant (stippled) and intolerant (white) native fishes in the five main rivers; the areas of the circles are proportional to the number of fish species, given below each river's name. ⊥⊥⊥⊥⊥, southern limit of ice at glacial maximum. ⌇⌇⌇⌇⌇, Continental Divide. FC, Fraser Canyon; FSJ, Fort Saint John, site of Charlie Lake Cave; PS, Puget Sound.

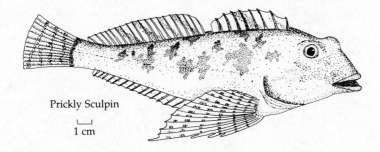

Prickly Sculpin

1 cm

water, migration north from the Columbia was more difficult. Presumably because of this, the number of strictly freshwater species in each river falls off markedly from the Columbia north to the Stikine, as shown by the pie graphs in the figure.

The five species of Pacific salmon, pink, chum, coho, sockeye, and chinook, alternate between salt water (where they live most of their lives) and fresh (where they spawn), and all five have successfully invaded the once glaciated rivers. In the spawning season, salmon no doubt provided abundant food for Paleo-Indians, whose remains[30] have been found on the banks of the Fraser River at Yale (see fig. 11.4).

The three-spined stickleback is also a salt-tolerant species and is found along both the Atlantic and Pacific shores of the continent. Some stocks spend part of their lives in salt water; others are restricted to fresh water. Taking all stocks together, their range is shown in figure 3.2. In the west, they are found in all the four once glaciated river systems listed previously, as well as in the Columbia.

Another salt-tolerant species—it lives mostly in fresh water but sometimes in salt water near river mouths—is the prickly sculpin. Its range extends much farther inland than that of the three-spined stickleback. Whereas the stickleback occurs in the Fraser River only downstream of the Fraser Canyon, the sculpin's range extends up to the headwaters of the Fraser and thence into the Parsnip River, a tributary of the Peace (itself a tributary of the Mackenzie) on the other side of the Continental Divide. It is believed to have crossed the divide in the following way.[31] The melting of Cordilleran ice was not uninterrupted: late in the process, a temporary readvance of the ice dammed the upper reaches of the Fraser, creating a lake that backed up right across the divide to flow away east, via the Parsnip River. The sculpins in the lake were transported passively, from a river system draining into the Pacific into one draining into the Arctic Ocean.

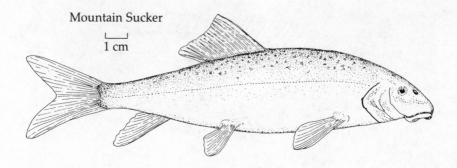

Mountain Sucker

⌊_⌋
1 cm

An example of a species that *cannot* tolerate salt water is the mountain sucker. Like three-spined sticklebacks, mountain suckers survived the glaciation in more than one refugium; in the suckers' case, the refugia were the Missouri River system to the east of the Continental Divide and the Columbia River system to the west. In the west, they managed to reach the Fraser from the Columbia, and they must have used a freshwater route because of their intolerance of salt water. There is no such route now, but from time to time during the melting of the Cordilleran ice, proglacial lakes formed in the several long, narrow, north-south valleys of the western mountains. At first these drained southward, into the Columbia; later, the receding ice front uncovered previously ice-filled valleys leading into the Fraser, whereupon the lakes overflowed northward. Many species of fish probably reached the Fraser from the Columbia through these reversing lakes. Their geography changed continually, providing different routes for different species. It is believed[32] that mountain suckers made the journey through the Arrow Lakes (see fig. 11.4).

Beringia at the Turn of the Epoch

As elsewhere, so also in Beringia, the vegetation changed around the time of the Pleistocene/Holocene transition—strictly speaking, somewhat before the transition, at about 11k B.P. Climatic warming is believed to have caused the change, as it had the earlier change at 14k B.P., when herb tundra was replaced by shrub birch tundra (see chapter 10). At 11k B.P., poplars and willow began to flourish throughout what is now the forested part of Alaska and the Yukon. Probably they had been there in small numbers all along, and when climatic change brought warm, dry summers, they underwent a population explosion[33]; this would explain why the increase in poplars and willows happened at the same time over a large area.

There was also a time-transgressive change: a fairly slow invasion, from refugia in the east, of spruce and alder. In easternmost Beringia, spruce began to increase at about 11k B.P., but it did not reach the westernmost limit of its present-day range, on the shores of Bering Strait, until 5k B.P.[34] Long before the slow migrants had finished their journey, however, the whole of interior Beringia pre-

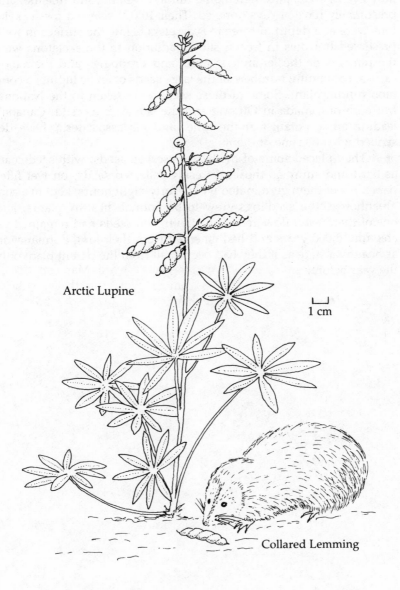

Arctic Lupine

1 cm

Collared Lemming

sented a totally different landscape; from being a treeless land of endless exposed barrens (tundra), it became a patchwork of forest (much of it muskeg), parkland, and tundra.

One of the surviving tundra patches, near the northern end of the modern Yukon-Alaska border, was the home of some collared lemmings, whose fossil remains have been found. These lemmings now live in arctic and high alpine tundra, well beyond treeline, and presumably have always done so. Their 10,000-year-old fossil skeletons were at a depth of three to six meters below the surface in well-preserved burrows in frozen silt. In addition to the skeletons were the remains of the lemmings' nests and droppings and their food caches, containing numbers of the large seeds of arctic lupine, a common tundra plant. Some of these seeds were taken to the National Museum of Canada in Ottawa, and the late A. E. Porsild, Canada's leading arctic botanist at the time, and his associates,[35] have described what became of them.

They placed some of the best-preserved seeds, with seed coats as hard and shiny as those of freshly collected seeds, on wet filter paper. Six of them germinated within forty-eight hours; kept in a cool greenhouse, the seedlings grew to be normal, healthy plants, and one of them even flowered. Thus the frozen seeds had remained viable for 10,000 years and had emerged from their long dormancy in as healthy a state as if they had been shed from the parent plant only the year before.

12

The Great Wave of Extinctions

The most striking ecological change marking the end of the Pleistocene epoch in North America was, sad to say, a great loss. In the space of three millennia at most, in the interval from 12k B.P. to 9k B.P., between thirty-five and forty species of large mammals became extinct.[1] This wave of extinctions is one of the most noteworthy, and most puzzling, events in ecological history. The reasons for it have been debated for decades, and none of the many explanations put forward is entirely satisfactory.

It is interesting to look at this tremendous extinction episode in its context in time and space. There are several questions to consider: Were there earlier extinction "waves," and what happened between waves? What happened in the rest of the world, outside North America? What kinds of animals became extinct? And if large mammals vanished, what about small ones? Were they involved too?

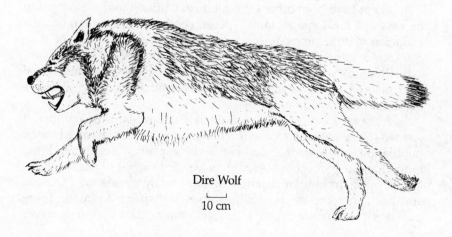

Dire Wolf

10 cm

To take the last question first: the term *large mammal* is unscientifically vague and needs to be defined. Paleontologists use a variety of definitions. Taking weight as the criterion, some[2] put the dividing line between "large" and "small" at forty-four kilograms; others,[3] at five kilograms.

The different definitions do not create as much confusion as might be thought; an intuitive classification puts the great majority into the right slot whichever criterion is used. Thus mastodons, mammoths, sabertooths, bison, shruboxen, muskoxen, camels, horses, bears, giant ground sloths, wolves, and the like are undeniably large. Mice, rats, squirrels, weasels, and the like are small. Surprisingly few are borderline.

In the wave of extinctions we are considering, in the last three millennia of the Pleistocene, most of the victims were large; only about five small victims are known.[4]

Extinction Waves: When, Where, and What

Extinction waves, or extinction episodes as they are often called, have happened many times in this continent's history. For example, at least six are believed to have occurred in the past ten million years, and they seem to have happened at the end of glaciations.[5] "Our" wave, the most recent of them, ranked second in the six, being exceeded only by one that occurred five million years ago. Indeed, considering only large mammals, ours was probably the worst. Although the total number of extinctions was slightly greater in the earlier episode, a much larger proportion of the victims were small species. (It is well to avoid giving precise numbers; new discoveries mean that the numbers change continually.)

There have been other extinctions in addition to those occurring in waves. A few large mammal species (perhaps nine or ten) became extinct at various times during the 60,000 to 80,000 years of the Wisconsin glaciation before the final 3,000 years. Indeed, in the long run, extinctions of species are as inevitable as the deaths of individual animals, and it may be that the causes of extinctions are as varied as the causes of individual deaths.

A wave of extinctions—a sudden diminution in the number of species—is analogous to a sudden big drop in the size of a human population, an event that deserves to be explained even though the individual people would inevitably have died sooner or later anyway. Catastrophes in human populations have many causes: war, famine, and pestilence are the possibilities that first spring to mind. There may be equally many causes for evolutionary catastrophes, as waves of extinctions could well be called. Another possibility, however, is that extinctions come in waves that are part of a recurring cycle. It would then be the cycle itself, rather than each individual wave in the cycle, that would need to be explained. If there is such a cycle, it presumably follows a cycle in the inorganic world, such as cyclical climatic changes.

Besides considering other times, we must also consider other places. Was the late Pleistocene extinction episode worldwide in scope, or was it confined to North America? Paleontological research shows that there were similar sudden waves of extinctions in South America and Australia. The South American wave came at the same time as the North American, but the Australian was 10,000 year earlier.[6] There were extinctions, too, in Eurasia and Africa, but in these two continents the losses were much less severe and they were spread over a longer period of time.

Thus the extinction wave seems to have involved the whole world. Even so, the differences in timing and duration in different parts of the world make it unlikely that the extinctions were caused astronomically. An astronomical cause has indeed been suggested, at least for the demise of the mammoths.[7] According to this theory, the earth picked up, temporarily, a diffuse, reflective shell of cometary particles above the atmosphere; as a result, sunlight was reflected, temperature dropped, and mammoths died out. But if this theory were correct, the extinctions would have been synchronous everywhere, and they were not.

Regardless of whether all the extinction waves of the past ten million years had the same cause or whether each was unique, it seems reasonable to suppose that all the extinctions *within* a wave had the same cause. The disappearance of thirty-five to forty large mammal species in the most recent wave can hardly be coincidence. It is worth inquiring whether the victims, as a group, simultaneously resembled one another and differed from extant large mammals. The answer seems to be no. The mammals that became extinct range from species almost indistinguishable from extant species to the totally unfamiliar.

To take some North American examples: the extinct dire wolf was a heavily built version of the extant timber wolf and had much bigger teeth.[8] The two species shared large parts of the continent, with dire wolves being much the more numerous until about 10k B.P. Then dire wolves became extinct, and timber wolf populations grew; thereafter, timber wolves were abundant until they were all but exterminated by modern humans. Other extinct species that closely resembled their living relatives were five species of Pleistocene horses

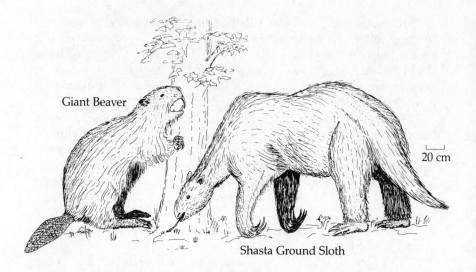

Giant Beaver

20 cm

Shasta Ground Sloth

and the western camel, all of which became extinct in North America between 12k and 10k B.P. They must have looked very like their extant counterparts (though modern camels differ enough to be assigned to a different genus), which do not live in North America.

Many of the species that became extinct differed strikingly from any animal now alive; they would be totally unfamiliar to us if it were not for museum restorations. Many have been mentioned on earlier pages. Among the most bizarre of the others were the giant beaver, as big as a black bear and with enormous incisors, and the Shasta ground sloth. It would be easy to extend the list.

All these now extinct animals either migrated into recently glaciated areas before extinction overtook them, or (presumably) would have done so if they had survived. The extinctions occurred throughout the ice-free part of the continent, and a variety of theories have been put forward to account for them. Indeed, the wave of extinctions is the subject of one of the great debates of ice age paleoecology.

There are two chief theories, each containing variant subtheories. According to one theory, the species that vanished were exterminated by human hunters; according to the other, the species that vanished were those unable to adapt to the rapidly changing environment. Let us look at the theories in turn.

The Prehistoric Overkill Hypothesis

The hypothesis that numerous species, in all continents, were driven to extinction by human hunting has become known as the prehistoric overkill hypothesis. Its chief proponent is Paul S. Martin of the University of Arizona.[9]

Considering the world as a whole, the supporting arguments are strong. Extinctions were most numerous, and most sudden, in those continents (North and South America, Australia) that were invaded by humans who had evolved and developed their hunting skills elsewhere. Extinctions were fewer and took place more gradually in Africa and Eurasia, the continents of human origin, where the earliest members of our species slowly evolved in lands they shared with herds of large mammals. It is reasonable to suppose that African and Eurasian mammals, living for generation after generation in an environment where human predators were one of the risks to be faced regularly, grew wary; the prey mammals adapted their behavior as fast (almost) as the hunters improved their skills. American and Australian mammals, by contrast, were unprepared for the onslaught when newly arrived, weapon-bearing humans began to hunt them; they succumbed without ever having time to develop evasive tactics.

It is believed that the numerous North American extinctions at the very end of the Pleistocene were the work of Clovis people (see chapter 5), who, judging from archaeological discoveries, came suddenly into prominence at 11.5k B.P.[10] Their ancestors presumably invaded from Beringia and penetrated to the lands south of the ice sheets as soon as the ice-free corridor became passable. Compared with the humans already living in midlatitude North America—if, indeed, there were any—the immigrant Clovis people were more numerous and more advanced in their hunting methods. They hunted many large mammals to extinction: in a word, their hunting amounted to overkill. According to this theory, the end of the Wisconsin glaciation was only indirectly responsible for the extinctions; it caused the opening of the ice-free corridor, the portal through which the hunters came.

There is further support for the overkill hypothesis. It was noted previously that the Australian extinctions were earlier than the North American. Correspondingly, humans arrived in Australia earlier than in midlatitude North America. However, there is not a perfect match between the two continents in the way effect (extinctions) followed cause (the arrival of human hunters). In North America, the extinctions seem to have started as soon as Clovis people arrived. In Australia, there was a gap[11]: the human population dates from about 40k B.P., whereas the wave of extinctions (which occupied 11,000 years) did not begin until 26k B.P.

Yet another argument in favor of overkill comes from comparing the mammals that did and did not go extinct in midlatitude North

America.[12] Twelve genera of grazers and browsers that no doubt served as human food disappeared between 11.5k B.P. and the end of the Pleistocene. They were camels, llamas, two genera of deer, two genera of pronghorn, stag-moose, shrub-oxen, woodland muskoxen, mastodons, mammoths, and also horses, which became extinct in North America though not elsewhere. All these animals were descended from ancestors that had lived in North America for more than one million years in an environment devoid of ruthless, expert human hunters (perhaps devoid of humans altogether). They never evolved the art of coexisting with so relentless a predator.

Now compare them with the nine genera of large grazers and browsers that were present in the Pleistocene and survive to the present day. They are bison, moose, wapiti or elk, caribou, deer, pronghorn, muskox, bighorn sheep, and mountain goat. All except pronghorn were immigrants from Asia; no doubt they were adapted to the presence of human hunters and able to survive in spite of being hunted.

It seems, indeed, entirely plausible that the two groups of animals—the native American victims and the immigrant survivors—differed in their capacity to withstand armed hunters. But notice that this argument is not equivalent to saying, simply, that immigrant spe-

Four-horned Pronghorn
⊢——⊣
10 cm

cies were wary and native American species too trusting. As present-day hunters know, existing game animals (all belonging to the nine genera just listed) may be wary or trusting, depending on whether they live in an area where hunting is permitted or excluded. The valuable adaptation possessed only by the immigrant species is the ability to learn from experience. This adaptation was the outcome of natural selection: animals that did not learn to be wary of human hunters have left no descendants.

There is no reason to believe that mammoths and mastodons, to take only the largest victims, were too big for human hunters to kill. They were probably as vulnerable to primitive hunting methods as African elephants are today. There are many ways of killing large prey.[13] They could have been captured in pitfalls, or with footsnares, and then been speared when they had become exhausted, using Clovis points as spear tips. (A footsnare is a device, weighted with a log, that clings to the leg of an animal that puts its foot into it.) They could have been caught in noose-type snares. They could have been killed by weighted spears suspended from tree branches. They could have been stampeded over cliffs, as was done in historic times with bison, which were driven over so-called buffalo jumps. It has even been suggested that they may have been hunted with poison-dipped spears, although this seems unlikely in North America, where few, if any, suitable plant poisons are available.

All in all, the arguments for human overkill as the cause of the extinctions seem, at first, very persuasive. But there are equally persuasive arguments against it.

The Arguments against Overkill

The chief objection to the overkill hypothesis is that, in Clovis times, the human population was small and human hunters few, too few in the opinion of many archaeologists to have had any significant effect on other animal species.

Another obvious objection is this. A species on its way to extinction is bound to become increasingly rare and hard to find as its numbers dwindle. Why, then, did human hunters not switch their efforts to more abundant animals, such as bison, whenever an earlier quarry became scarce? (Perhaps this objection can be overruled by arguing that when a particular quarry became so scarce as to be not worth hunting, it was doomed to dwindle to extinction anyway.[14])

The same objection applies to nonhuman predators, both extant (such as the timber wolf) and extinct (such as the sabertooth). It seems at first thought reasonable to suppose that the extinct carni-

American Lion

10 cm

vores died out because their food supply vanished. But why did they not, like surviving predators, switch to surviving prey? A possible explanation is suggested at the end of the following section.

It is worth comparing the origins of extant and extinct carnivores in the same way as we compared the origins of extant and extinct herbivores in the preceding section, even though the comparison does not suggest any conclusions. The facts, so far as they are known, are these.

Five species of carnivorous mammals disappeared at the end of the Pleistocene (enough is known for us to consider species rather than genera). They were the giant short-faced bear and the American cheetah, which may both have become extinct before 11.5k B.P., when Clovis people became numerous; the American lion; the sabertooth; and the dire wolf. The lion is believed to have been a fairly recent immigrant from Asia; the origin of the dire wolf is unknown; and the other three descended from North American ancestors.

Only four species of large, "menacing" carnivores have survived. They are the timber wolf, the grizzly bear, the cougar, and the wolverine. (Because they are rarely a threat to large herbivores, we exclude coyotes, black bears, badgers, red and gray foxes, lynxes and bobcats, even though all are extant large carnivores if the division between large and small is put at five kilograms. Polar bears are excluded, too, because they are marine.) Grizzly bears are quite recent immigrants to middle latitudes; they arrived (as did moose) from Beringia after the ice sheets melted. Timber wolves and wolverines have North American ancestors. And cougars have been here at least 200,000 years.

Still another compelling argument against the overkill hypothesis is that fossils of extinct mammals are hardly ever found in association with human remains at archaeological "kill sites." The exceptions to this generalization are mammoths and extinct species of bison. However, the genus *Bison* still survives; it was not killed off during the late Pleistocene extinctions; rather, the species of bison living then have been replaced by modern descendants. As for mammoths, their bones have been found at about one-third of the known archaeological sites representing the Clovis culture,[15] and they often show signs of butchering. The majority of these places are in the southern United States. The map in figure 12.1 shows most of the known Clovis sites in land within or near the southernmost margins of the last ice sheets; sites with mammoth fossils are marked, as are sites with remains of other extinct species. These last are clearly rare. One might infer that human hunters were selective and for an unknown reason concentrated nearly all their efforts on bison, mammoth, and animals that happen to be still extant, such as caribou and elk.

There is another possibility, however. The proponents of the overkill theory have a counterargument to explain why evidence of hu-

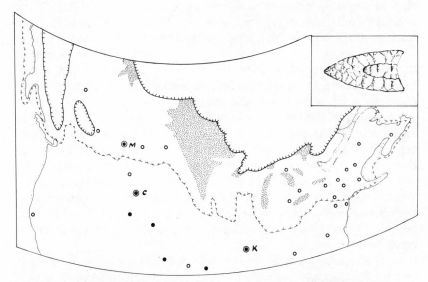

FIGURE 12.1: Midlatitude (thirty-five to sixty degrees north) North America, showing ice sheets, lakes, and coastline at 10k B.P., and archaeological sites dating from the Clovis period (11.5k to 10k B.P.). O, sites with signs of human occupancy only. ●, sites with human and mammoth remains. ◉, sites with remains of humans and other extinct mammals: M, Medicine Hat, Alberta (mammoth, Mexican wild ass, camel?); C, Colby, Wyoming (mammoth, camel); K, Kimmswick, Missouri (mastodon). The ice front at 18k B.P. is also shown. Inset, a Clovis point.

man life and the fossils of extinct mammals are so seldom found together. It is as follows.[16] It assumes that an invading horde of human hunters, of unprecedented ferocity, entered midlatitude North America through the bottleneck formed by the ice-free corridor and spread in all directions, occupying an ever expanding, more or less circular area. The periphery of this area formed a narrow front, along which the hunters encountered defenceless North American game animals for the first time. They annihilated them. As the area occupied by the ever growing human population expanded, the front moved continuously outward. Therefore, hunters and hunted were simultaneously present only in a narrow, moving zone, which crossed any given spot of ground in a very short time (a few decades or centuries).

The theory seems farfetched, but it gives a possible explanation of why fossils of the extinct animals (other than mammoths) are so seldom found at archaeological sites. Fossils of any kind are rare, and those that have been found represent the accumulated deaths of tens of thousands of years. If the hypothesis is correct, only deaths occurring in the final decades of a species' existence would have been caused by human hunting, and these deaths amounted to only a tiny fraction of the accumulated total. All earlier deaths would have been natural, leaving fossils in "natural" deposits (that is, deposits devoid of human artifacts). The theory does not explain why mammoth remains are comparatively common at archaeological sites, whereas mastodons, horses, camels, and the like are so rare.

Although the first version of the theory assumed that the geographical starting point of the invading hunters was in the ice-free corridor, according to a later version[17] the invasion came from northeastern Siberia and was responsible for the Beringian extinctions as well as those south of the ice. This cannot be reconciled with the arguments[18] of prehistorians who maintain that Clovis culture, and with it the skills and tools needed to hunt large animals efficiently, developed south of the great ice sheets among primitive people already living there.

This brings us back to the unsolved problem of when people first reached the lands south of the ice. If overkill could only be perpetrated by people of Clovis culture, and the ancient Beringians were too primitive, the extinctions in Beringia are left unexplained. It is even possible that human invaders were for a time prevented from crossing the Beringian land bridge by the larger, fiercer carnivores already there. Human hunters pursued their prey in a dangerous environment. They, as well as the animals they hunted, were prey

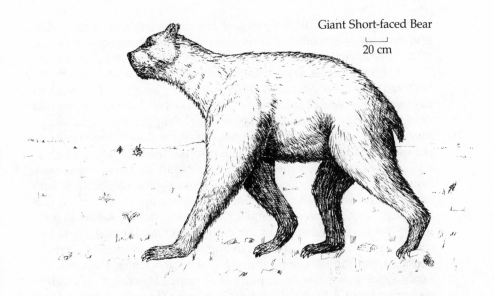

Giant Short-faced Bear

20 cm

themselves, and no doubt many people were killed and eaten by American lions and the terrifyingly big, swift, agile, and ferocious short-faced bears.

Changing Environment Theories

Most paleoecologists who disbelieve the overkill theory invoke environmental change, resulting from the climatic change that brought an end to the Wisconsin glaciation, as the only possible cause of the great wave of extinctions. There are a number of such theories. Some fail because they apply to only one or two animal species, others because they apply only in small regions. It strains credibility to suppose that different species, in different places, all went extinct for different causes at roughly the same time.

General theories, designed to explain all the late Pleistocene extinctions everywhere in North America, face a very serious objection even before they are propounded. Surely the climatic amelioration and the tremendous expansion of habitable land area that came with the disappearance of the ice sheets should have caused animal populations to grow bigger rather than dwindle to extinction. The phenomenon to be explained is the direct opposite of what common sense would suggest. We must therefore inquire: How can seemingly better conditions kill whole populations of many large mammal species?

Here are some of the suggestions that have been made. Most lead only to "particular" theories, that is theories explaining the extinction of a single species or several extinctions in a single locality.

The extinctions in Beringia seem easy to explain for those who believe that at the height of the Wisconsin glaciation the area was covered by so-called arctic steppe, which provided copious forage for herds of grazing animals (see chapter 7). Disappearance of the steppe must have caused disappearance of the herds. The steppe began to shrink soon after Bering Strait broke through the land bridge at 15.5k B.P.; as the sea flooded the extensive Beringian lowlands, the climate became milder and moister, and the grassy arctic steppe (if it ever existed) gave way to shrub tundra and later to forests of poplar, willow, alders, and spruce. The major vegetation change seems to have happened long in advance of the extinctions, but if one accepts that the disappearance of suitable fodder caused mammoths, camels, horses, long-horned bison, and the rest to go extinct, then one can easily argue that a few small patches of steppe lasted long enough to accommodate a few shrinking herds of game for as long as the fossil evidence requires. However, the theory is unconvincing to disbelievers in the arctic steppe, and it does not apply to extinctions south of the ice sheets.

Several other particular theories are based on environmental change, or habitat destruction, as it could equally well be called. For example, it may have been habitat destruction that drove mastodons to extinction.[19] Between 12k and 10k B.P., their favorite browse—spruce—disappeared from the big region south of the Great Lakes where mastodons were most numerous. Spruce forest was replaced by pine forest, which was acceptable if not optimal for them. But then pine, too, became scarce, because of the northward advance of deciduous forests. The pine forests became fragmented into isolated patches that became smaller and smaller. Deprived of a habitat that had been at least tolerable, the mastodons died out, probably by 10k B.P.

Similarly in the west, the changing climate in the rain shadow of the Rocky Mountains caused a change in the available fodder. The increasing aridity of the plains caused short grasses to replace taller grasses. This, it is argued,[20] put grazers adapted to comparatively tall grasses, such as horses, camels, and mammoths, which can digest grass stems, at a competitive disadvantage vis à vis bison, which graze on the leafier short grasses. Consequently horses, camels, and mammoths disappeared, and bison flourished in their stead.

Loss of habitat has also been blamed[21] for the extinction of the giant beaver, which, like modern beavers, inhabited ponds and swamps. Competition with modern beavers for the shrinking supply of suitable habitat may have contributed to their extinction, but it seems impossible to judge why *Castor* should have defeated *Castoroides* in the struggle for living space rather than the other way around.

Likewise, competition has been blamed for the extinction of the huge stag-moose, which was probably a muskeg dweller. It is argued that when the melting ice allowed the similar but smaller modern moose to migrate into midlatitude North America from Beringia, stag-moose was the loser in the competition for habitat. It is not clear, however, why the invader should have succeeded in crowding out the established species.

However, to say that a species became extinct because of loss of habitat entails a circular argument. With the disappearance of the ice sheets, *all* habitats changed; and when a habitat changes, the old habitat can be described either as "changed" or "destroyed." It is a matter of semantics. It is easy to say that the habitat of an extinct species was destroyed, whereas that of a still extant species was merely changed. But if the reason for saying so is simply that the extinct species is extinct, and the extant one extant, then the statement is not an explanation at all, but a play on words.

Many animals—all the survivors—managed to adapt to the changes that accompanied disappearance of the ice sheets. Some responded by altering their geographical ranges. A good example[22] is provided by four small mammals that currently inhabit markedly different habitats: the smoky shrew of mature, eastern deciduous forests; the thirteen-lined ground squirrel of the prairies; the heather vole of the boreal forest; and the Ungava lemming of the tundra. Nowadays their ranges do not overlap (fig. 12.2), but fossils of all four species, dating from 11k B.P., have been found together in a cave in Pennsylvania. There is no way of knowing how extensive their shared range was then.

The ancient environment at the fossil site was probably an open parkland in which spruces, pines, and birch grew amidst ground vegetation having some of the attributes of tundra and some of prairie. Such a mixed environment, which has no modern counterpart, may have covered a large area, and it would have met the requirements of all four species. For the thirteen-lined ground squirrel, for instance, it would have served as an eastern extension of the western grasslands (see chapter 11). When the mixed environment was destroyed, instead of going extinct the animals fanned out into the four separate regions where their different requirements could be met. One wonders why all animals whose habitats were changed did not

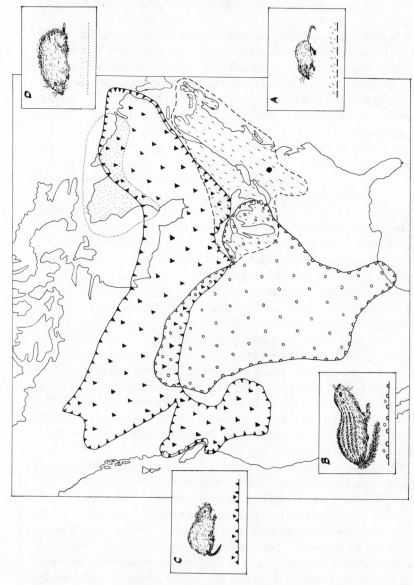

FIGURE 12.2: The present-day ranges of (A) the smoky shrew, (B) the thirteen-lined ground squirrel, (C) the heather vole, and (D) the Ungava lemming: ●, site in Pennsylvania where 11,000-year-old fossils of all four species have been found together.

do likewise. Some paleoecologists argue that environmental change is much less damaging to small mammals than to large; presumably the reason for saying so is that very few small mammals went extinct! Some theories assert that the changed climate itself caused the extinctions directly, independently of changes in the vegetation. According to one of these theories,[23] the opening of the ice-free corridor between the Laurentide and Cordilleran ice sheets created a funnel for bitterly cold winds out of the arctic; the result is believed to have been a catastrophic fall in temperature that killed off the horses, camels, and bison living at the southern end of the corridor.

Another "climatic" theory[24] argues that at the end of the Pleistocene, the climate became much more seasonal that it had been; summers were hotter, winters colder, and the seasons were more strongly contrasted in their precipitation as well. Hence, according to the argument, the period during each year in which conditions were mild enough for the survival of newborn young would have become shorter. Large herbivores may have had a reproductive cycle (the interval from one birth to the next) that failed to synchronize with the seasonal cycle. Many births would have taken place in unfavorable times of the year, and extinctions were the result.

The numerous "environmental" theories put forward to account for the extinctions—the ones described are only a sample—all fail (in my opinion) in being too farfetched or too "particular" (in the sense previously defined). Moreover, they all seem to overlook the fact that tremendous environmental changes occurred during the Wisconsin glaciation as well as at the end of it. The overkill theory has fatal objections, too.

Could it be that some short-lived catastrophe killed off vast numbers of all large mammals (or all large herbivores) and that species now extant are those few that managed to build up their numbers again after the catastrophe was over? If so, what *was* the catastrophe? It would have had to have been one that left no evidence of its occurrence and was short lived enough to leave no perceptible gap in the fossil record. All that can now be said is that the cause of the great mammal extinctions is still an unsolved puzzle.

Extinct Birds

The great wave of late Pleistocene extinctions affected birds as well as mammals.[25] The casualties were mostly flesh eaters, especially carrion feeders: extinct species of eagles, vultures and condors, and teratorns.

These birds probably relied on the dead bodies of large herbiv-

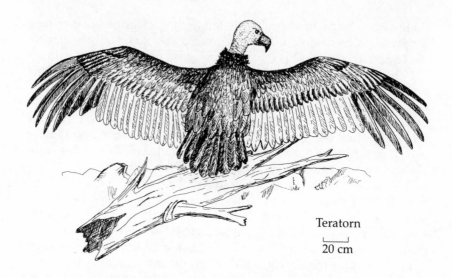

Teratorn

20 cm

orous mammals for a constant supply of carrion. They were as dependent on them as were the mammalian carnivores—sabertooths, American lions, dire wolves, and the like. It therefore seems reasonable, at first thought, to blame the extinctions of all the flesh eaters, birds as well as mammals, on the disappearance of the herbivores.

Against this theory it could be argued that even when mammoths, mastodons, camels, horses, and all the other now extinct herbivores vanished, the flesh eaters had only to switch to bison, elk, caribou, deer, and all the other living herbivores to survive. But, perhaps, as already suggested, some undiscovered catastrophe caused all species of herbivores to be scarce for a short period. Then all flesh eaters—mammals and birds alike—would have been at risk of starvation. Though many species died out, a few of the tough ones survived, and their descendants are with us still.

If there was indeed a catastrophe, its nature has not even been surmised. It is worth repeating that the great wave of extinctions at the end of the Pleistocene has yet to be convincingly explained.

PART FIVE

OUR PRESENT EPOCH, THE HOLOCENE

13

The Great Warmth

In the past 10,000 years (or in the Holocene epoch thus far), the North American climate completed its postglacial warming and started on the long cooling trend leading to the next glaciation. The moment at which temperatures stopped drifting upward and began drifting downward varied from place to place. The moment in the Milankovitch cycle when summer sunlight reached its greatest intensity in high northern latitudes was about 10,000 years ago (see chapter 1), a time when the great ice sheets were still enormous (see fig. 1.4).

These huge masses of ice melted very slowly, and the last mainland remnants of the Laurentide ice sheet, to the east of Hudson Bay, did not disappear until 6.5k B.P. or even later.[1] The climate at any point on the ground thus resulted from two opposing influences: the warmth of the sun, still great even after it had begun its slow decline, and the cooling effect of nearby ice. The moment of greatest warmth at any point depended, therefore, on its geographical location. Places in the far west experienced their "climatic optimum" as much as 6,000 years earlier than places close to the last surviving patch of ice in Labrador. The plains just east of the Rocky Mountains experienced warm, dry summers (warmer and drier than they are now) while much of Quebec and Labrador were still under an ice sheet hundreds of meters thick. The map in figure 13.1 shows a highly speculative reconstruction of the way the wave of maximum warmth swept across the continent. The isolines join places that experienced their optima synchronously.

From the ecological point of view the moment of maximum warmth is merely the midpoint of a warm period. Nearly everywhere in glaciated North America there has been a period since the end of the last glaciation during which the climate was appreciably warmer than it is now, and usually much drier as well. This period has no

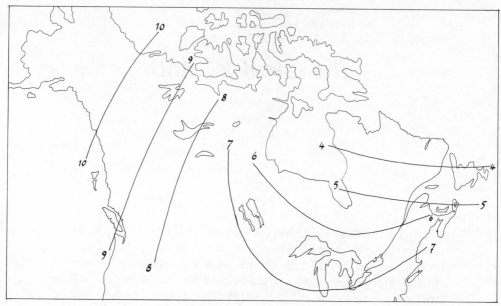

FIGURE 13.1: The isolines join points where the climatic optimum occurred at the same time. The times are given in thousands of years B.P. (The map is highly speculative.)

identifiable beginning and end; temperatures gradually rose to their highest level and then, as gradually, fell. Nevertheless, the interval of time in which temperatures were above present-day averages was a "warm" period as judged by present-day standards and, as such, has been given a variety of names: the hypsithermal; the altithermal; the xerothermic (applied where the climate was dry as well as warm); and the climatic optimum (a term more appropriate to an instant of time than to an interval). The first of these names is now used almost universally.

In this chapter, we consider what the land and the sea were like during the hypsithermal. Were conditions much as they are now except that, because of the greater warmth, all biogeographical zones were shifted tens or hundreds of kilometers northward? Or would the scenery have been altogether strange to modern eyes?

The answer is probably yes to both questions. In most places the hypsithermal lasted for three or four millennia, long enough for tremendous ecological changes to take place. Conditions at the end must have been altogether different from conditions at the beginning. Also, the course of events varied greatly from one place to another.

In some regions, for instance near the southern limits of the Laurentide ice sheet, the land emerged from its icy cover while the climate
was still cool and warmed very gradually. But near the center of the
Laurentide sheet, the land's final emergence from rapidly melting ice
(often after a number of false starts blotted out by subsequent ice
surges) was straight into summer sunlight stronger than today's.

Where emergence was rapid, the combination of warm climate
and undeveloped soil must, at first, have allowed some unfamiliar
ecosystems to develop on terrain such as that found now only in
the high arctic. Moraines deposited by the ice would have formed a
jumble of steep, unstable, rapidly eroding hills; among them flowed
torrential meltwater rivers. Outwash deposited by the rivers formed
mosaics of different textures, ranging from boulders to the finest
rock-flour clay. A combination of what we now think of as temperate
zone vegetation and high arctic landforms would have produced scenery unlike anything known today.

Where newly ice-free land was suddenly exposed to warmth
and dryness, conditions were probably harsh, though not in the
sense of being cold. The proximity of warm land to cold ice produced
a steep temperature gradient and, consequently, strong winds. Continual gales must have swept across the country before vegetation
had developed to act as a brake. As long as there were no plants to
diminish its force at ground level, the wind picked up quantities of
loose dust, sand, and grit from the quickly drying till, producing dust
storms that darkened the sky for weeks at a time. Great tracts of sand
dunes were built up in some places, and in others thick layers of
loess.

In spite of the odds against it, vegetation slowly became established, and the winds abated. Evidence about the winds comes from
sediment cores collected in Lake Superior.[2] The grains forming the
postglacial sediments become smaller the less deeply they are buried;
they range in size from coarse sand at the bottom, deposited about
9.5k B.P., to fine clay above, at the 6.5k B.P. level. Coarse sand implies
strong currents in the lake, which in turn imply strong winds; conversely, fine clay implies weak currents moved by gentle breezes. The
changing texture of the sediments in a sediment core thus gives a
picture of the changing scenery on land, from windswept desert at
and before the beginning of the hypsithermal to forest by its close.

A snapshot of conditions late in the hypsithermal in the Great
Lakes region is given by fossil finds along the north shore of Lake
Ontario[3] that date from between 5k and 6k B.P. They suggest a moist,
rich forest of beech, sugar maple, and eastern white pine with grassy
openings. The diversified habitat was home to a variety of small

Redwings and Wood Duck

mammals—chipmunks, squirrels, flying squirrels, deer mice, meadow voles, and shrews and, a step up in the food chain, red and gray foxes. There were also ponds and lakes inhabited by muskrats, wood ducks, and red-winged blackbirds. The contrast with conditions a few millennia earlier could hardly be greater.

Some Northward Shifts of Northern Limits

Fossil finds show that the geographical ranges of numerous plant and animal species extended farther north in the hypsithermal than they do today. These fossils provide the evidence that there was, indeed, a warm period in the past and show also how the warmth slowly drifted from west to east across the continent. Here are some examples. (The locations of all the places mentioned in this chapter are shown in the map in figure 13.2.)

The earliest signs of warmth, possibly dating from a little before 10k B.P., come from the Seward Peninsula, Alaska, in what was unglaciated Beringia. Fossils of beaver dams and beaver-gnawed wood have been found, as well as fossil logs of poplar, birch, and spruce.[4] The peninsula is beyond the modern range of beavers, birch, and spruce, and only a few small poplars remain where big ones once flourished. The large size of the fossil logs shows that the trees grew in deep soil; permafrost was probably much less widespread than it is at present.

Indeed, the soil as well as the atmosphere warmed up during the hypsithermal. Probably much of the subarctic where permafrost now occurs was then free of it, while in higher latitudes the top of the permafrost was at a greater depth below the surface than it has been before or since.

In the far northwest of the Northwest Territories the northern limit of forests in the early Holocene is known to have been some distance north of its present position. During the warm period, which probably lasted from about 10k to 6k B.P., much of what is now tundra was forested. An example is the Tuktoyaktuk Peninsula; at about 10k B.P. it was invaded by spruce, and the large quantities of pollen in lake sediment cores show that the vegetation must have been a true forest of spruce, not merely scattered trees. Cattails and sweet gale evidently grew in wetlands in the forest; both were beyond their modern northern limit, which for cattails is several hundred kilometers to the south. When the warm period ended, the forest disappeared, to be replaced by tundra that still remains.[5]

Other tree species besides spruce grew north of their present northern limits; these northern advances happened right across the continent, one after another as each region in turn experienced its warm spell.[6] In Vancouver Island, on the west coast, Douglas-fir grew farther north around 9k B.P. than it does now.[7] North of Lake Superior, eastern white pine reached a limit about 150 to 200 kilometers north of its modern limit between 6.5k and 6k B.P.[8] Investigations[9] in tundra near Dubawnt Lake (west of Hudson Bay) have revealed buried soil covered by a layer of charcoal fragments—the remains of a forest fire. The charcoal, and hence the fire, date from 3.5k B.P.; if

Beaver Dam

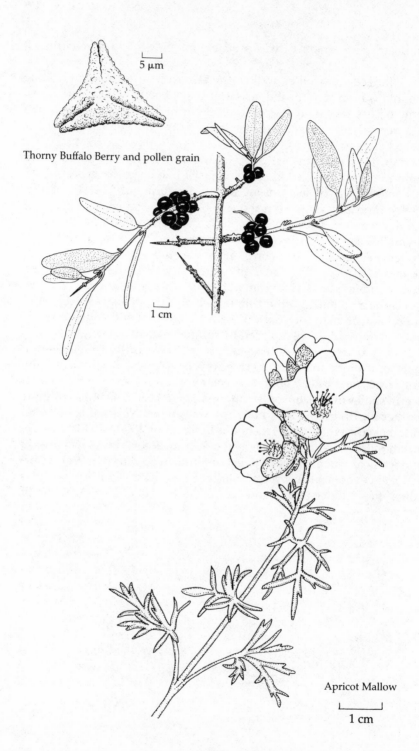

5 μm

Thorny Buffalo Berry and pollen grain

1 cm

Apricot Mallow

1 cm

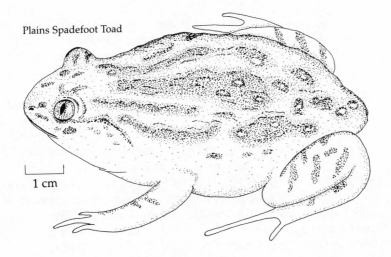

Plains Spadefoot Toad

1 cm

there was a forest fire, there must have been a forest, at a site now 280 kilometers north of the tree line. In Labrador, the evidence that big trees once grew where now the land bears a parkland of much smaller trees or tundra consists of massive subfossil spruce logs eroding out of peat beds.[10]

As the boreal forest shifted northward, so did the prairie immediately south of it. From about 10k to 6k B.P., Riding Mountain, Manitoba, now a forested national park, was evidently part of the prairie; pollen evidence shows that such typical prairie plants as apricot mallow, moss phlox, and thorny buffalo berry grew there, from which it has been inferred[11] that average summer temperatures were about two degrees Celsius higher than at present, and annual precipitation about three centimeters less than the current forty-five centimeters. At the height of the hypsithermal, the prairie-forest border lay north of Porcupine Mountain, Manitoba, more than 100 kilometers north of its present position.[12] Because of the dryness, the prairie spread eastward as well as northward, to about 120 kilometers beyond its present eastern limit.[13] At first the trees were replaced by prairie grasses; then, as the drought worsened, ragweed, sagebrush, orache, and pigweed replaced some of the grasses. At times there may have been severe dust bowl conditions.

Aquatic plants also extended their ranges. For example,[14] a fossil of coontail was found east of Great Slave Lake at a north latitude of sixty-four degrees; it is not now found north of sixty degrees north latitude.

Animals as well as plants, of course, lived farther north than at present during the hypsithermal. For instance, a fossil skeleton of the plains spadefoot toad, an animal of the dry shortgrass prairie, was found at Killam, Alberta, more than 100 kilometers north of its present limit.[15]

Examples could be multiplied indefinitely. Insects, especially beetles, deserve mention. Because of their structure, they always leave an abundance of undecayed remains. Consider some data from Kitchener, Ontario, consisting of insect fragments and pollen in a core collected from lake sediments laid down between the time the ice first disappeared (soon after 13k B.P.) and 7k B.P.[16] The microfossils in the core show how plant and insect communities changed in response to the warming climate until, by 10.5k B.P., the insects were very much as they are in the region today. But temperatures continued to climb for at least 2,000 years more; by 8.5k B.P. the insect community resembled the kind of community now found hundreds of kilometers to the south in the east-central United States. For example, a species of rove beetle was present that is not now found north of southern Ohio, between 300 and 400 kilometers to the south.

Rove Beetle (Homaeotarsus)

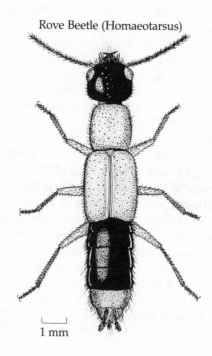

1 mm

The Hypsithermal at Sea

The climatic warming of the hypsithermal affected the sea as well as the land. Evidence from the high arctic shows that the Arctic Ocean was less extensively frozen in the summer during the period 6k to 4k B.P. than it was before and after this period. The evidence comes from Ellesmere Island, the northernmost island in the Canadian Arctic Archipelago, and it consists of stranded driftwood logs.

Given sufficient warming of the northern climate, summers become warm enough to melt much of the sea ice in the Arctic, especially the landfast ice along shorelines. This permits driftwood logs floating in Arctic Ocean currents to become stranded on beaches where dry land meets the sea; such strandings cannot happen if a zone of landfast ice separates land and water. Driftwood logs are fairly plentiful in the Arctic Ocean; they are carried down into it by the rivers of western Canada (especially the Mackenzie) and Siberia. Therefore, if old logs are found on ancient, raised beaches, it follows that when the logs were deposited, the beaches must have been ice free in summer. Fortunately, raised beaches are common in the Arctic islands; this is because the islands have been rebounding ever since the ice sheets began to melt. The combination of raised beaches and stranded logs, whose ages can be determined by radiocarbon dating, has made it possible to reconstruct the Holocene history of the Arctic Ocean[17]; logs ranging in age from 4,000 to 6,000 years are especially numerous. Hence the conclusion that this period was the arctic hypsithermal.

At lower latitudes, the warming ocean allowed many marine organisms to increase their ranges northward. For example, between 3k and 6k B.P., eastern oysters lived on the shores of Sable Island; they are not found there now.[18] The island was larger at the time as sea level was probably between twenty-five and thirty meters below its present level at 7k B.P. (see The East Coast Plains and Islands, chapter 6).

Many other shore organisms were able to enlarge their ranges northward.[19] Not only did the increased warmth of the summer sun (due to the Milankovitch cycle) raise the temperature of the sea; as well, the cold Labrador current, which nowadays brings icy Arctic Ocean water southward as far as Cape Cod, was deflected eastward by submarine sandbanks, especially the Grand Banks of Newfoundland (see fig. 6.5), which were less deeply submerged than they are now.

Because of these two warming influences, summer sea-surface temperatures at the height of the hypsithermal were probably greater

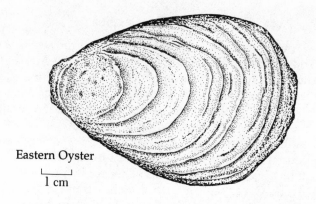

Eastern Oyster

1 cm

than eighteen degrees Celsius as far north as Cape Breton Island and into the Gulf of Saint Lawrence; nowadays they are less than fifteen degrees Celsius from about Bangor, Maine, northward, except in the enclosed southwestern embayment of the Gulf of Saint Lawrence, around Prince Edward Island. There, summer warmth continues; it has not decreased appreciably since the hypsithermal, despite the general climatic cooling, because the water is too shallow to be affected by deep-water currents. Consequently, a variety of organisms, such as some sand fleas, a mud crab, and a marine snail, now have disjunct ranges; they are found along the Atlantic shore from Maine southward and in the southwestern Gulf of Saint Lawrence, but not between. Presumably their ranges were uninterrupted at the height of the hypsithermal. Then, when the climate cooled, they disappeared from exposed shores north of Maine, surviving only in the unusually protected waters around Prince Edward Island.

The Hypsithermal in the Mountains

At the same time as the warmth of the hypsithermal was causing the tree line and other ecotones (ecological boundaries) to shift northward, in mountainous country it was causing them to shift upward. Evidence comes from the mountain ranges in both eastern and western North America.

In the east, consider an example from the White Mountains of New Hampshire, where sediment cores were collected from six lakes at different elevations. The pollen and conifer needles in the sediments show that between 9k and 5k B.P. eastern white pine and eastern hemlock grew at elevations of 300 or 400 meters above their present elevational limit.[20] Assuming that the lapse rate (the rate at which temperature falls with increasing elevation) was the same in the past as it is now, this implies that mean annual temperatures in the region must have been about two degrees Celsius higher than at present. At present, too, the mean annual precipitation is greater the higher the elevation; equivalently, precipitation increases as temperature falls. From this it follows that if (and it is a big *if*) the relation between precipitation and temperature was the same in the past as it is now, the mean annual precipitation at all elevations would have been 400 millimeters less than at present. Even if the numbers are inexact, at least it seems reasonable to conclude that conditions during the hyp-

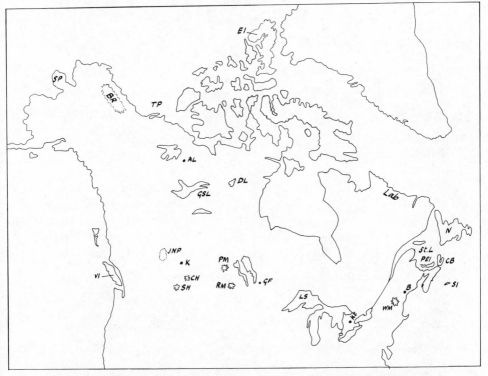

FIGURE 13.2: Places named in chapter 13, listed alphabetically. AL, Acasta Lake; B, Bangor, Maine; BR, Brooks Range, Alaska; CB, Cape Breton; CH, Cypress Hills; DL, Dubawnt Lake; EI, Ellesmere Island; GF, Great Falls, Manitoba; GSL, Great Slave Lake; JNP, Jasper National Park; K, Killam, Alberta; Kt, Kitchener, Ontario; Lab, Labrador; LS, Lake Superior; N, Newfoundland; PEI, Prince Edward Island; PM, Porcupine Mountain; RM, Riding Mountain; SH, Sweetgrass Hills; SI, Sable Island; SP, Seward Peninsula; St. L, Gulf of Saint Lawrence; TP, Tuktoyaktuk Peninsula; VI, Vancouver Island; WM, White Mountains, New Hampshire.

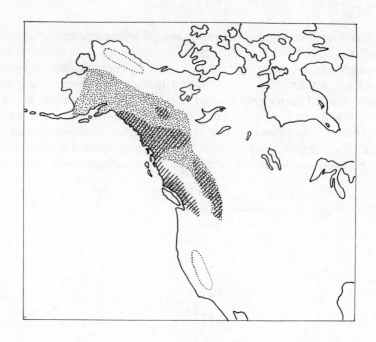

FIGURE 13.3: The present geographical ranges of mountain goat (left) and hoary marmot (right). Dotted outlines show the Brooks Range, Alaska, and the Sierra Nevada, California.

sithermal were appreciably drier, as well as warmer, than they are now.

Now consider the west, specifically the Rocky Mountains. Studies in Jasper National Park in Alberta have shown that by 9k B.P., or soon after, the tree line was at least 100 meters above its present level.[21] The evidence comes from the discovery of large fir and spruce logs buried in peat bogs of the alpine tundra above the modern tree line and also from the needles and pollen in sediment cores. The pollen data suggest that tree line was high from 8.8k to 5.2k B.P., except for a century-long interruption beginning about 7.4k, when it descended, temporarily, to its present level. No doubt this indicates a small climatic "wobble;" climate is never constant, and there is no reason why the hypsithermal warmth should have been continuous.

The dates of these events were estimated in the usual way, by radiocarbon dating of the peat in sediment cores. In addition, a useful control date was available. Jasper National Park is within the fallout area of Mazama ash, the volcanic ash produced by the eruption of Mount Mazama in Oregon at 6.7k B.P. Sediment cores taken in the region contain a layer of this buried ash, and it serves as a reference level, or stratigraphic marker

The buried logs have made it possible to infer past temperatures by a method known as *isotope dendrochronology*.[22] Atmospheric oxygen consists of a mixture of different oxygen isotopes (variants of oxygen with different atomic weights); the commonest isotope, forming 99.76 percent of the total, is ^{16}O (with atomic weight 16); the next commonest, forming about 0.2 percent, is ^{18}O (with atomic weight 18). It is known that the ratio of ^{18}O to ^{16}O in rainwater, and hence in the tissues of living plants, varies with temperature. Therefore, by determining this ratio for samples of ancient wood, it is possible to deduce the prevailing mean annual temperature at the time the wood was formed. Applying the method to the logs from Jasper National Park shows that temperatures were at least 0.5 degrees Celsius higher than at present around 8.5k B.P. in the first half of the hypsithermal and about 1.5 degrees Celsius higher around 5.5k B.P. in the second half, after the century-long cool period.

The upward shift of vegetation zones in the western mountains during the hypsithermal brought about changes in the areas of the different zones. As the upper limit of tree growth ascended the slopes, the patches of alpine tundra at high elevations dwindled. This is believed to account for the modern geographical ranges of two familiar mountain mammals, the mountain goat and the hoary marmot.[23] Their present ranges are shown in figure 13.3. Fossil finds show that both species survived the glacial maximum south of the

Richardson's Ground Squirrel

⊢⊣
1 cm

ice; indeed, they ranged farther south than they do at present; there were mountain goats as far south as the Sierra Nevada. It is unlikely that they lived north of the ice as well, in Beringia; had they done so, they would probably be found in the Brooks Range of Alaska to this day, but neither species ranges that far north; nor do they live in Siberia. The evidence therefore suggests that they spread into their present ranges from the south when the Cordilleran ice sheets melted. Subsequently, as the fossil evidence shows, they died out in some places, presumably where their required habitat—alpine tundra—dwindled and disappeared in the warmth of the hypsithermal. This may explain the present-day absence of the two species from the southern Rockies and Coast-Cascade ranges.

While the ranges of mountain goats and hoary marmots are thought to have contracted because of the warmth, the ranges of some lowland mammals expanded. For example, consider Richardson's ground squirrel, at present a common species of the dry western grasslands, both in the Great Plains (east of the Rocky Mountains) and in the Great Basin (between the Rockies and the Sierra Nevada). These two parts of its range are now disjunct. The species is thought[24] to have survived the glaciation in the Great Basin and to have spread eastward and northward when grassland replaced the spruce forests that had covered the lands south of the ice sheets at glacial maximum. To have crossed the Rockies, however, these prairie-dwelling ground squirrels would have had to migrate over mountain passes that are now heavily forested. The obvious explanation for their present range is that, during the hypsithermal, the boundary between grassland and forest shifted upward to a higher elevation, bringing some of the passes that are now forested into the grassland zone where Richardson's ground squirrels can thrive.

Refugia from the Drought

As we have seen in earlier sections, the warmth and dryness of the hypsithermal caused forest and grassland to shift northward all across the western half of the continent and also upward wherever there were hills and mountains. In the Great Plains, in the rain shadow of the Rocky Mountains, the steadily increasing drought was especially marked. Some hills in the midst of the plains were high enough, however, to intercept a few moisture-bearing clouds; consequently, the rising forest-grassland boundary never reached their summits, even though they were (and are) well south of the present latitudinal position of that boundary. Forests still remain on the tops of these hills, forming islands in a great sea of grassland (fig. 13.4).

Ever since the hypsithermal these hilltops have provided refugia for forest ecosystems unable to endure the heat and drought (especially the drought) of the plains. They are *dry climate refugia* in the same way that nunataks were glacial refugia. They account for many disjunctions in the geographical ranges of plants.

Figure 13.5 shows two such plants: heart-leaved arnica and thimbleberry.[25] At present, these two species are common forest plants at low elevations in the western mountains, and they are also found at a few points in the forests around the Great Lakes. But in the grasslands between, they have managed to survive only where hilltop refugia enabled them to escape the drought of the plains, for example, on the Cypress Hills of Alberta and Saskatchewan, the Sweetgrass Hills of Montana, and the Black Hills of South Dakota.

The list of refugees varies from one refugium to another. Even refugia as close together as the Cypress Hills and the Sweetgrass Hills—they are separated by a mere 150 kilometers—have strikingly different forests.[26] Thus the only coniferous trees in the Cypress Hills are lodgepole pine and white spruce. The Sweetgrass Hills support a much greater variety: Douglas-fir, alpine fir, limber, whitebark, and lodgepole pines, and hybrid spruce. The Sweetgrass Hills' hybrid spruce trees are descendants of crosses between the white spruce of eastern North America and the Engelmann spruce of the western mountains.

The contrasts between the two refugia raise at least three puzzling questions: First, why should the mixture of trees in the Cypress Hills consist of one eastern and one western tree, whose modern ranges overlap nowhere else? Nowadays lodgepole pine is a tree of the western mountains, whereas white spruce is found only east of the continental divide. Second, why should the spruces of the Cypress Hills be "pure" white spruce while those of the Sweetgrass

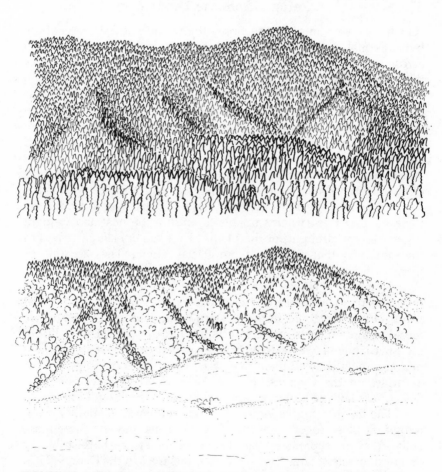

FIGURE 13.4: The Cypress Hills of Saskatchewan before (above) and after (below) grassland replaced the postglacial forests covering the plains.

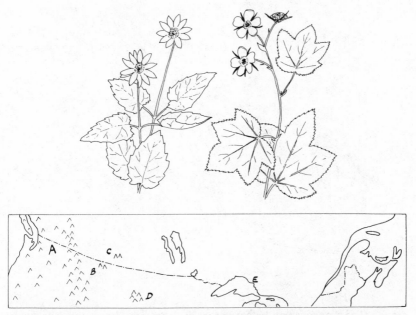

FIGURE 13.5: Two plants whose ranges probably became disjunct in the hypsithermal. Left, Heart-leaved arnica. Right, Thimbleberry. The disjunct areas where they now grow are marked on the map with letters. A, The western mountains; B, The Sweetgrass Hills of Montana; C, The Cypress Hills of Alberta and Saskatchewan, D, The Black Hills of South Dakota; E, The shores of the Great Lakes.

Hills are hybrids between white spruce and Engelmann spruce? (Engelmann spruce, like lodgepole pine, is a tree of the western mountains.) Finally, why should there be so many more species in the Sweetgrass Hills than in the Cypress Hills?

Only the last of these questions seems to have a straightforward answer. The Sweetgrass Hills are higher than the Cypress Hills (the respective elevations are 2,100 meters and 1,400 meters). Possibly the trees now absent from the Cypress Hills did grow there before the hypsithermal reached its peak but were then "driven off the summits" by the increasing heat and drought; presumably the higher Sweetgrass summits remained damp enough and cool enough for the trees to persist.

The isolated hilltop refugia in the Great Plains are the sites of whole disjunct ecosystems. In addition to the trees, each contains a host of forest organisms that require the shady habitat created by trees; they probably migrated from the plains into the hills with the

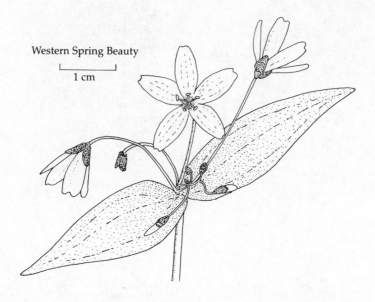

Western Spring Beauty

1 cm

trees. (It is necessary to say *probably* because these are inferences based on modern geographical ranges.) The Cypress Hills forests, for example, provide an island habitat for numerous herbaceous plants typical of coniferous forests, for example, western spring beauty, Calypso orchid, pipsissewa, bunchberry, twinflower, and many more.[27]

Animals of various kinds, too, share the refugia and have disjunct ranges as a result; for example, two species of shrew (the dwarf shrew and the northern water shrew) and the heather vole are disjunct in the Sweetgrass Hills.[28] In the Cypress Hills are tiger salamanders, the red-sided garter snake,[29] and nine species of stonefly[30] that require cool, moist habitats.

Human Life in the Hypsithermal

Among the large mammals that escaped the wave of extinctions at the end of the Pleistocene was, of course, our own species. The human population of glaciated North America slowly increased, and a variety of modes of life developed, each adapted to its own ecosystem and changing with it as the climate warmed. There were maritime cultures around the coasts and hunting cultures with different specialties in the interior: bison hunters in the grasslands and caribou hunters in the tundra and forest-tundra of the north.

These people experienced the hypsithermal. They enjoyed (though that may be the wrong word) a warmer climate than we do

now. For arctic peoples the warmth was a mixed blessing, or no blessing at all, since it shortened the season during which landfast sea ice
provided a safe surface for traveling and hunting. Farther south, in
the forests and plains, frequent forest fires and grass fires and severe
drought also made life more difficult in some ways than it had been.

The peoples of the coasts evidently became capable maritime
hunters.[31] In what had been Beringia, they may have been forced to
turn to the sea for subsistence because, when forest invaded the hitherto open country of the interior, such grazing animals as woolly
mammoth, horses, camels, bison, and wapiti (elk) either became extinct or migrated away in concert with their migrating food plants.

On the Pacific coast, along the shores of British Columbia and
the Gulf of Alaska, the people are believed to have successfully
hunted large marine mammals: whales, seals, walruses, and Steller's
sea cows. The last-named were the only sirenians adapted to cold

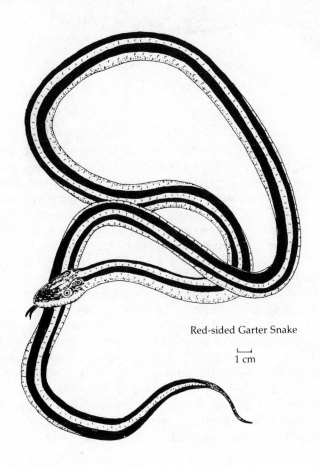

Red-sided Garter Snake

1 cm

Steller's Sea Cow

waters; they were large (about five tons), tame, slow-moving vegetarians that grazed on kelp close to the shore. Their vulnerability doomed them, and by 200 years B.P. they had been hunted to extinction.[32] However, to kill animals better able to defend themselves than sea cows, the human hunters must have constructed seaworthy boats of whale skin or seal skin, possibly kayaks with watertight decking. And they also made the necessary weapons, probably lances at first and, later, harpoons with detachable heads. Besides hunting sea mammals, they no doubt fished for halibut, cod, and salmon and gathered oysters and clams on the beaches.

This maritime modus vivendi is believed to have been fully developed all along the Pacific coast of once glaciated North America by about 8k B.P., even though along the southern part of the coast there are, at present, few archaeological sites dating from before 4.5k B.P. The lack of sites is believed to be only apparent; almost certainly they are there but have been submerged by the rising sea and will be revealed in due course, when underwater archaeologists search for them.

The last part of the continent to experience the hypsithermal was Labrador, where the final remnant of the Laurentide ice sheet persisted in the interior until 6.5k B.P. or possibly even later, and the coast was (and still is) cooled by the frigid Labrador current. Even so, Labrador's Atlantic coast was populated by humans from very early in the Holocene, in the north by Paleo-Inuit (Paleo-Eskimo) people and farther south by Paleo-Indians. Archaeological evidence suggests[33] that the boundary between the two cultures shifted, first northward as the climate warmed and then, after the peak of the hypsithermal at about 4k B.P., southward again. The cause seems to have been that during the warming period, the Paleo-Indians migrated northward in concert with the shifting boundaries of vegetation zones, but even so they may have relied for subsistence on the sea as much as the land. Since few of the archaeological sites contain any animal remains, it is hard to judge.

The vast interior of the continent was also populated, rather sparsely, by human hunters. Those living between about 10k and 7.5k B.P. are described as belonging to the Plano culture; this culture followed two earlier cultures, the Clovis (see chapter 5), which lasted from 11.5k to 11k B.P., and the Lindenmeir (or Folsom), which lasted from 11k to 10k B.P.[34] Whereas the two earlier stone age peoples used "fluted points" as weapons, the Plano people made narrow, leaf-shaped points and later constructed "stemmed points," with a stem at the base for attachment to a shaft. Plano people lived after the great wave of extinctions at the end of the Pleistocene. Bison had now become the chief quarry in the Great Plains because the great variety of game available in earlier millennia was gone. It is tempting to speculate on how many generations of human children marveled at stories of their forefathers' mammoth-hunting exploits and of their encounters with sabertooths; we shall never know.

Bison hunters probably used driving as their chief hunting method.[35] The reason for believing this is that most of the Paleo-Indian archaeological sites containing animal remains (mostly bison bones, horns, and teeth) contain them in abundance; the sites are "mass kill" sites. The hunters presumably drove large herds of bison into natural impoundments such as narrow canyons or into specially constructed corrals, where they could be slaughtered. In other places, the animals were killed or disabled by being driven over buffalo jumps. The hunters used natural landforms to guide the herds in the direction they wanted them to go. As one anthropologist has put it,[36] the bison drives amounted to "risky free range husbandry."

The gradual onset of the hypsithermal undermined this way of life in many places. Increasing heat and drought, no doubt accompanied by plagues of grasshoppers and devastating prairie fires, ravaged the grasslands. The bison herds were forced to retreat from large areas of the Great Plains in search of richer pastures, which they presumably found near the prairie-forest border, where the climate was wetter. A place where bison in their thousands may have been slaughtered is Great Falls, Manitoba (see fig. 13.2), where an archaeological site, dated at about 8k B.P., has yielded more than 50,000 Plano artifacts.[37] The site must have been occupied soon after it was left high and dry by the northward-migrating Lake Agassiz (see fig. 9.4). Available evidence shows that part of the site was used for killing, skinning, and butchering large animals, and the animals were probably bison. This is not completely certain, however, because no organic remains have been found; forest has replaced the earlier grassland, and bones, horns, hooves, and teeth have all been dissolved by the humic acids in the forest soil. Therefore, the animals could have been (though they probably were not) caribou.

North and east of the great grasslands, caribou was the most important game animal. The Plano people of the forests spread northward in the wake of the receding ice and depended for most of their needs on caribou, which, like the bison of the grasslands, provided meat, hides for clothes and tents, bones for utensils and tools, and sinews for thread.

Farther north, the open country of the subarctic would have been more hospitable for both caribou and their human pursuers than the dense, relatively unproductive spruce forest immediately adjacent to the grasslands. The closed forest gradually thins out in the north, into open woodland carpeted with "reindeer moss" (lichens, mostly of the genus *Cladonia*), the chief food of caribou; the lichen-woodland zone itself merges, farther north again, into treeless tundra. Barren-ground caribou no doubt migrated from the woodland to the tundra every spring and back into the woodland every fall, just as they do now. They were probably hunted by various methods: they may have been driven into impoundments, snared, captured in pitfalls, or speared as they swam across rivers.

Probably people dependent on caribou for their survival lived clear across the continent during the hypsithermal. Evidence for their presence comes from sites as far apart as Labrador and Acasta Lake[38] in the Northwest Territories (see fig. 13.2). The increasing warmth and dryness of the climate constituted a threat to all these people's welfare; forest fires became more frequent and more extensive, destroying the lichen on which the caribou depended. After a fire, especially one that burns the organic soil, lichens take a long time to regrow. Huge tracts of land must have become uninhabitable for caribou and therefore for humans. Thus the end of the hypsithermal must in some ways have been welcome, even though it marked the onset of the next glaciation.

14

The Neoglaciation

The Milankovitch cycle proceeds inexorably; we are now embarked on the next glaciation, the neoglaciation as it is called. It had no well-marked beginning. The changes in the earth's orbit around the sun which make up the Milankovitch cycle caused the start of a trend toward cooler summers at high northern latitudes as early as 9k or 10k B.P. But this cooling was masked, to begin with, by the warming that resulted from the slow disappearance of the continental ice sheets. The great mass of ice disappeared so slowly that the last of it melted away long after the astronomically determined moment of maximum summer warmth; melting could and did continue even after the onset of astronomically determined cooling. The neoglaciation can be said to have begun when the cooling caught up with the warming. The cooling was accompanied by increased precipitation.

In a nutshell, the North American climate has been getting colder and wetter since 4k or 5k B.P. But the trend has not been steady, since the Milankovitch cycle is not the only astronomical cause of climatic change. As noted in chapter 1, it is believed that the sun's output varies cyclically, with a period of 2,500 years; as a result, minor warmings and coolings are superimposed alternately on the slow, majestic Milankovitch cycle (see fig. 1.6). These lesser oscillations are sometimes called the *little ice age cycle*.[1] The clearest evidence for them comes from the alternate expansions and contractions of the glaciers in the western mountains. Most of the mountain glaciers melted in the hypsithermal. They are believed to have formed anew, shortly before 5k B.P., when solar cooling added to Milankovitch cooling brought an end to warmth in the west. Studies[2] of moraine geomorphology show that since then, in response to the solar fluctuations, the mountain glaciers have retreated and advanced twice, with the advances peaking at about 2.8k and 0.3k B.P. Between these cold periods were respites—temporary ameliorations in the neoglaciation.

The ecological effects of these respites are described later in this chapter. First we consider the dominant effects of the neoglaciation, of which one of the most marked was the spread of muskeg.

The Spread of Muskeg

The remarkable vegetation known as muskeg was described in chapter 4. It is otherwise known as peatland, or sometimes simply as bog. It develops on poorly drained land wherever the climate is wet and cool: wet enough for a continuous layer of peat mosses and feather mosses to cover the ground and for fires to be rare; and cool enough for decomposition to take place so slowly that dead moss and other plant remains accumulate undecayed, as peat. It is the continuous accumulation of undecayed plant material together with the water absorbed in it that distinguishes muskeg from other kinds of freshwater wetlands such as sedge marshes (fens) or forested marshes (swamps), in which the water is not so tenaciously held by the dead plants.

Some botanists[3] distinguish between *muskeg* and *bog*. The two types of vegetation develop differently: muskeg forms directly on flat, poorly drained land, whereas a bog results from the infilling of what was once a pond or lake of open water. Although the results are indistinguishable, cores from the peat show the difference. In "true" muskeg, the peat is moss peat from top to bottom; in a bog the peat is in layers, with limnic peat (compressed lake plankton and algae) and sedge peat at the bottom, overlain by moss peat. In what follows, the distinction will be ignored.

Tracts of muskeg are common, nowadays, throughout the boreal forest. Nonbotanists are blind to its charms. One critic[4] said that it is "smeared across Canada like leprosy . . . a rotting mushland of blackflies and mosquitoes." Botanists know it for its abundance of attractive bog plants: leatherleaf, sweet gale, bog rosemary, Labrador tea, swamp laurel, cotton grass, and sundew, to name a few. Where the ground is truly waterlogged, the only trees are black spruce and tamarack.

Although quantities of muskeg no doubt developed on the newly ice-free land uncovered by the shrinking ice sheets, much of this early muskeg dried up and disappeared in the warm, dry hypsithermal climate. It reestablished itself and expanded when cool, moist conditions returned, with the start of the neoglaciation. For

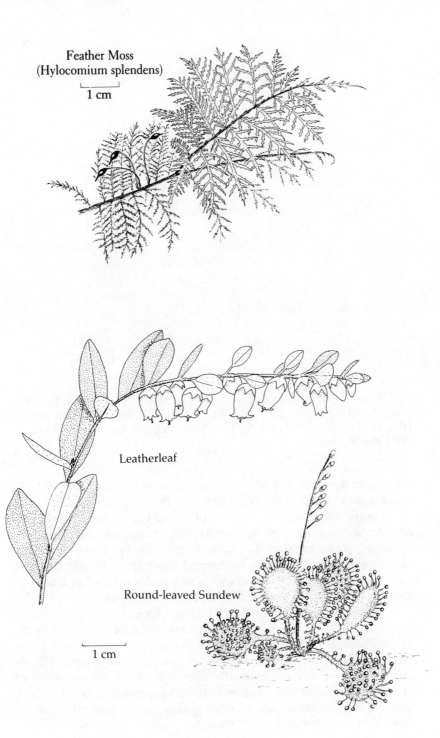

Feather Moss
(Hylocomium splendens)

1 cm

Leatherleaf

Round-leaved Sundew

1 cm

293

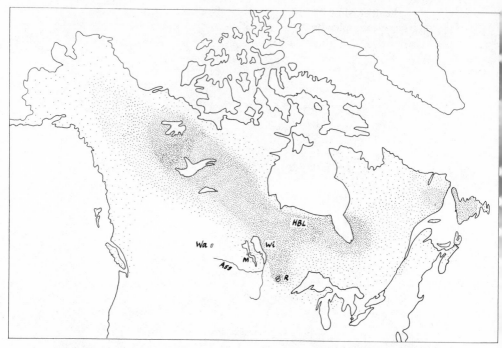

FIGURE 14.1: The modern distribution of muskegs. The density of the stippling shows their frequency. Wi, Lake Winnipeg; R, Red Lake, Minnesota; HBL, Hudson Bay Lowlands; M, Lake Manitoba; Ass, Assiniboine River; Wa, Lake Waldsea.

muskeg to develop properly, the ground must be constantly wet. Because it was so flat and poorly drained, the lake bed left behind by the disappearance of Lake Agassiz provided big tracts of suitable terrain, notably in a strip along the east side of modern Lake Winnipeg stretching south as far as Red Lake, Minnesota (fig. 14.1). Here muskeg developed over previously formed swamps and fens[5]; these, in turn, were the respective successors of woodlands and prairies that had flourished in the warm, dry hypsithermal climate.

Tracts of muskeg are even more numerous in the area known as the Hudson Bay Lowlands. This is the area that was drowned by the Tyrrell Sea (see chapter 10) when the Laurentide ice sheet first melted, leaving the earth's crust still depressed by the weight of the ice. As the land surface gradually rose, because of slow isostatic rebound, the Tyrrell Sea emptied. It left expanses of level, poorly drained marine sediments, an ideal terrain for muskeg, which developed in abundance and is still there.[6]

Muskeg has also grown up over much of the area where permafrost prevails and has presumably spread southward, following the permafrost, as neoglaciation has progressed. Permafrost ensures poor drainage and waterlogged soil during the growing season even where precipitation is low. Meltwater in the active layer (the soil that thaws each summer) is prevented from draining downward by the impervious frozen ground beneath; thus permafrost creates the conditions for muskeg. A feedback situation can then (sometimes) develop: the peat formed by the muskeg acts as a thermal insulator preventing the ground below it from thawing; thus muskeg preserves the permafrost that allowed it to grow in the first place.

However, it could happen, in some places, that an insulating blanket of muskeg, already growing on unfrozen ground, might prevent permafrost from advancing. In this case the peat layer would protect the underlying soil from freezing rather than from thawing. Obviously, the relationship between permafrost and muskeg is not at all straightforward; what happens on any patch of ground depends on the interactions, over a long stretch of time, of changing climate and changing vegetation.

Patches of muskeg are also to be found in the coastal forests of northern British Columbia[7] and the Alaska panhandle.[8] The cool climate and heavy rains provide the right conditions, and muskeg would probably be much commoner than it is if it were not for the rugged topography; much of the land is too steep. The muskeg patches are believed to have developed with the onset of the neoglaciation. Western muskeg resembles eastern muskeg ecologically, even though many of the plant species are different. The most striking difference is in the trees; in place of the tamarack and black spruce of eastern muskeg are stunted, scrubby shore pines (a subspecies of lodgepole pine) and yellow cedars.

Muskeg is also common on the eastern side of the continent, especially in Newfoundland. Indeed, patches of muskeg are found scattered throughout the northern forests, though with varying frequency, as figure 14.1 shows. North of the muskeg belt the climate, though cold, is too dry for muskeg, and south of the belt either too warm, too dry, or both.

The replacement of "ordinary" forest, in which dead plants decay and surplus water runs off, by muskeg, in which neither of these things happen, is known as *paludification*. It amounts to a deterioration or degradation of the vegetation in the sense that plant growth, and indeed the whole rate of materials cycling, is slowed. Paludification thus seems to be one of the ways in which life slows down when a new glaciation starts.

Increased Rain in the Prairies

The dry prairie grasslands, stretching from Lake Winnipeg to the Rockies south of the boreal forest, were not unaffected, of course, by the neoglaciation. They had expanded during the hypsithermal; now they contracted again, especially in the east, where the prairie-forest border retreated westward 120 kilometers from its farthest east position. As elsewhere, the climate became cooler and wetter, and although the remaining grasslands were still too dry for trees, the ground vegetation became less desertlike; sagebrush, orache, and cactuses gave way to grasses, except at the very driest sites.

Prairie lakes and ponds must also have been affected by the increased precipitation and reduced evaporation rates of the neoglaciation, but the effects were to some extent masked by other factors. The land that had been weighed down by the Laurentide ice sheet, and that had then been submerged under a succession of ice front lakes, was still rebounding. Because the terrain is so nearly flat, even a very slight difference in the amount of rebound between one point and another could "warp" the topography and produce profound changes in the drainage pattern. An example is the Assiniboine River which, before 4.5k B.P., flowed from the prairies into the Red River south of Lake Winnipeg (see fig. 14.1); a change in topography then diverted the river into Lake Manitoba, another of the large existing remnants of Lake Agassiz.[9] The result was an increase in lake volume over and above that caused by increased precipitation. Subsequently, at about 2.2k B.P., the river reverted to its old route, bypassing Lake Manitoba; the lake level remained fairly constant, however, because the climate continued to get wetter and cooler.

Thus lakes become deeper and shallower for a variety of causes, not solely in response to climatic change. Increased drainage into a

Brine shrimp

└─────┘
1 mm

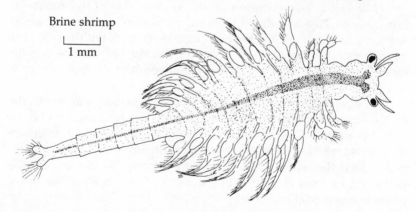

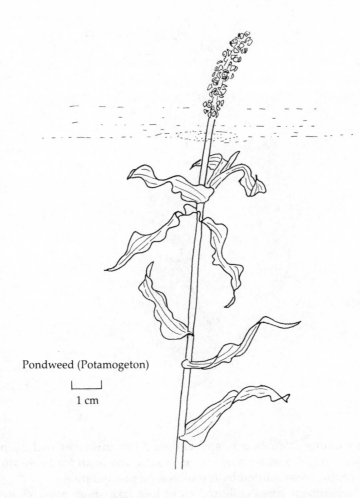

Pondweed (Potamogeton)

1 cm

lake can even make it shallower rather than deeper if increasing amounts of sediment are brought in and deposited on the lake bottom. Therefore the ecological history of a lake, as revealed by the fossils and microfossils in its sediments, cannot always be directly interpreted as a climatic history. This fact is treated as something to deplore by those who are interested in paleoecology only as an indicator of past climates.[10] But for a true paleoecologist, all ecological change is interesting, whatever its cause.

A lake whose increased depth probably was caused by the wetter climate of the neoglaciation is Lake Waldsea, near Humboldt, Saskatchewan (see fig. 14.1). Besides becoming deeper, it also became much less saline.[11] The evidence consists of the eggs of the brine shrimp, which are found only in the lowermost, oldest sediment

Stagnant Pond Snail

5 mm

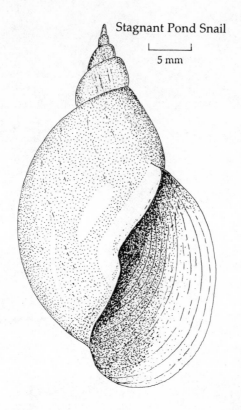

layer dating from 4k B.P. and earlier. After this layer had been laid down, the lake water must have become too fresh for brine shrimp; no doubt it was diluted by the increased precipitation.

Sediment cores collected close to a lake shore often show that there have been fluctuations in the water level without any change in water chemistry. For example, if a core collected from a cattail marsh contains fossils of submerged, deep-water plants, such as pond-weeds and water milfoil, it follows that the water was deeper in the past. Conversely, if a core from the middle of a lake contains shells of the stagnant pond snail, an inhabitant of wet shores, it follows that the lake used to be smaller. In neither case would it be possible to infer what had caused the change in water level.

Many lakes and ponds in the prairies show such signs of fluc-tuating water levels. The fluctuations may be due to changes in pre-cipitation or to isostatic adjustments of the earth's crust; but which-ever the immediate cause, the ultimate cause is the cycle of glaciations.

The Shifting Ranges of Forest Tree Species

The range limits of many species of trees shifted because of the neoglaciation. Or rather they continued to shift as the climate continued to cool down from its maximum warmth at the height of the hypsithermal. The range change of eastern white pine is representative of several species.[12] Eastern white pine retreated from its earlier "farthest north" extent (see chapter 13) toward its present northern limit (because of the increasing cold) while increasing its range westward into what had been prairie grassland (because of the increasing wetness). Figure 14.2 shows its modern range.

However, not all tree species "retreated" as the climate cooled. One that did not is hickory. Hickories continued to spread northward and eastward from their glacial refugium in the lower Mississippi Valley for a long time, perhaps 2,000 years, after the climatic optimum was past; they reached Connecticut at about 5k B.P.[13] They were among the slowest of tree migrants, so new territory remained ahead of them even after their potential northern limit, set by temperature, had begun to shift southward.

Another slow migrant was eastern hemlock; its slowness resulted, at least in part, from its need for organic seedbeds in which to establish itself, and these took a long time to develop. But that was not the only obstacle faced by the species. It is a good example of a tree whose fate was not governed solely by climate and soil, and a reminder that other vicissitudes must be allowed for too. Evidence from both pollen and macrofossils show that, at about 4.8k B.P., there was a sudden, massive dieback of eastern hemlock throughout the whole of its then geographical range.[14] The species remained scarce for a millennium and then recovered, partially, for another millennium; it is still much less abundant than it was before the crash. Because the disaster was so sudden and struck at the same moment over such a large area, climatic change is an unlikely cause. Either a sudden epidemic of some fungus disease or an outbreak of some serious insect pest is probably to blame. The most likely insect is the hemlock looper, still an important hemlock pest. However, the hemlock looper attacks balsam fir as readily as it does hemlock, and fir shows no detectable change in abundance at 4.8k B.P. The exact cause of the hemlock decline is probably undiscoverable.

A tree that spread far to the north during the neoglaciation was jack pine. Like eastern white pine, it survived the Wisconsin glaciation in the Appalachian Mountains. But unlike eastern white pine, which cannot endure minimum winter temperatures much below minus forty degrees Celsius, jack pine is adapted to survive temperatures as low as any that ever occur naturally (see chapter 8), and its northward migration into unglaciated territory was to a large extent unhindered by the increasing cold. It colonized areas covered by well-drained, sandy and gravelly glacial outwash, where muskeg could not develop. It is believed to have reached its present northernmost limit about 2.4k B.P.[15] This limit is probably set by the climate of summer; even when trees are not killed by extreme cold, they cannot reproduce successfully unless the summer growing period is long enough and warm enough.

Meanwhile, as described in chapter 4 and figure 4.3, the closely related lodgepole pine had been migrating northward from a southern refugium in the western mountains. The ranges of the two species now overlap (fig. 14.2), and within the area of overlap they hybridize freely. Lodgepole pine arrived in the overlap area first, at about 5.6k B.P., or 3,000 years ahead of jack pine.[16] (The history of this pair of tree species resembles that of the three pairs of related bird species described in chapter 3, but whereas the birds' migra-

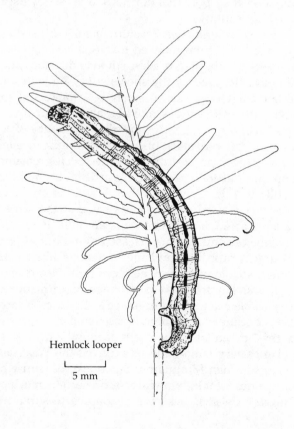

Hemlock looper

|_____|
5 mm

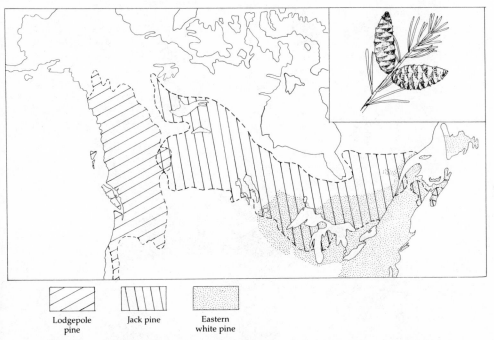

FIGURE 14.2: The modern ranges of eastern white pine, jack pine, and lodgepole pine. Where the ranges of jack pine and lodgepole pine overlap, they form hybrids. Inset, twig of a hybrid.

tional histories can be inferred only from their present geographical ranges and therefore cannot be dated, that of the trees is inferred from fossil pollen, which can be dated.)

The Neoglacial and the Northern Treeline

By the time the neoglaciation started, both white spruce and black spruce had advanced as far north as they could grow. White spruce, in particular, had occupied the newly ice-free land very rapidly (see chapter 4). Black spruce benefited from the neoglaciation to the extent that its preferred habitat—muskeg—expanded in area. Black spruce is one of the few trees (tamarack is another) whose seeds germinate on a seedbed of peat.

However, as the summers became cooler, the northernmost conifers (white and black spruce and tamarack) were unable to reproduce sexually because the season of summer warmth became too short for their seeds to mature. The climatic cooling did not kill the

A Black Spruce Clone

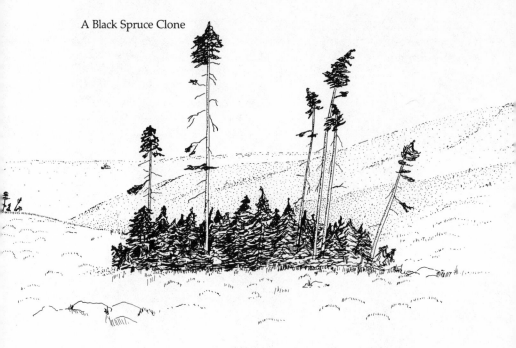

trees; like jack pine, they are adapted to withstand the coldest of arctic winters once they have become established. Even so, the northern limits of these species would have been forced southward as existing trees died of old age had it not been for the fact that they were able to propagate themselves asexually; this they do by *layering*. This mode of propagation produces a clone of trees, all genetically identical to one another. Black spruce layers easily; white spruce and tamarack, less so.

The result is that the two spruce species still survive successfully, as clones maintained by layering, well to the north of the line marking their limits of sexual reproduction. (Tamarack clones seem to be uncommon.) In subarctic Canada to the west of Hudson Bay, such relict clones of dwarf spruces often grow in sheltered sites in the tundra, far north of sexually reproducing members of their species.[17] Some of the clones may date from as long ago as the hypsithermal, making them 4,000 or 5,000 years old; others probably got started during one of the short-lived warm spells that punctuated the neoglaciation. If the clones were to be destroyed by fire, they would not be replaced, since no seed sources are nearby. But if the climatic cooling trend were to reverse itself, no doubt the cloned trees would eventually resume sexual reproduction.

These two spruce species each have, in fact, two "treelines": one beyond which they cannot now reproduce by seed, although they must have done so once or they could never have become established there in the first place, and a more northerly "clone line," beyond which seed production has always been impossible so that clones could never get started. In some places the southward retreat of the clone line during the neoglaciation can be inferred. Thus on the Arctic Ocean shore west of Hudson Bay, fragments of spruce wood have been found in sediments dating from 3.7k to 2.5k B.P.; there are no spruce remains after the later date and no spruce pollen in the sediment at all (except for minute amounts obviously blown in from far away). The wood fragments are evidently the remains of spruce clones that failed to produce pollen; now even the clones are gone, perhaps because the Inuit (Eskimos) used the wood for construction or as firewood.[18]

Only to the west of Hudson Bay do the spruces have both a true treeline and a clone line. East of the Bay, they reproduce by seed as far north as they grow, and there is no northern fringe of asexual clones.[19] The contrast has not been satisfactorily explained.

Refugia Reestablished

During the hypsithermal, it will be recalled, the vegetation zones on hills and mountains shifted upward; we noted in chapter 13 how the boundary between hot, dry grasslands and cooler, wetter forests moved upward on isolated hills in the western plains, turning the hilltops into islands of forest in a sea of grassland. Similarly in the high mountains: the forest-tundra boundary rose up the slopes (see The Hypsithermal in the Mountains, chapter 13; Refugia Near the Ice-free Corridor, chapter 7), leaving isolated islands of alpine tundra surrounded by montane forest on the summits.

With the coming of the neoglaciation, some of these tundra islands have become surrounded by newly formed ice fields formed by coalescing glaciers and so have been converted into nunataks. Probably many of the ice fields now in the western mountains (see fig. 1.1) are no older than 4,000 or 5,000 years; the ice that covered the same ground at the height of the Wisconsin glaciation melted in the hypsithermal, and the ice that exists there now is neoglacial ice. Consequently the summits protruding through it are neoglacial nunataks. Some of them may have been nunataks all through the Wisconsin; if so, a few of the plants growing on them now could conceivably be descended from ancestors that have grown at the same site for 100,000 years or more. And some of the new nunataks may continue

Purple Saxifrage

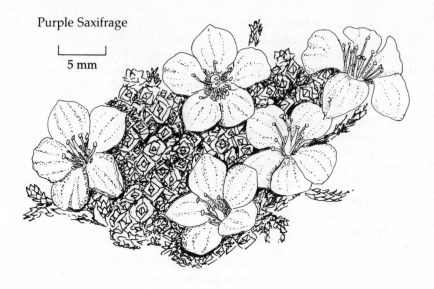

as nunataks through the whole of the coming glaciation, enabling a few plant lineages to survive from our interglacial to the next.

According to Calvin Heusser,[20] one of the foremost authorities on the glacial history of western North America, the majority of the existing nunataks in the Juneau ice field of southeastern Alaska are neoglacial nunataks. Even if the ice field did not melt completely in the hypsithermal (and it probably did), most of the summits that are now nunataks would not then have been surrounded by ice. An abundance of alpine plants grows on these nunataks now, plants such as purple saxifrage, sibbaldia, crowberry, mountain sorrel, mountain goldenrod, partridge foot, mountain fireweed, yellow heather, moss phlox, and many more. A few of them may possibly have survived in place, for generation after generation, through the whole of the Wisconsin glaciation, even though the majority probably moved up from the surrounding tundra-covered lowlands when the hypsithermal began. A few may have found their way in since the neoglaciation started, as the ice field fluctuated in response to the climatic oscillations of the little ice age cycle.

One other kind of habitat, in addition to nunataks, may persist unchanged through glaciations and interglacials—limestone caves. A well-studied example, Castleguard Cave in the Alberta Rocky Mountains, was described in chapter 7. There seems no reason why the two species of aquatic crustaceans that have (probably) occupied the cave system continuously since the last (Sangamon) interglacial should not go on living there. In time, the Columbia Icefield will expand over the roof of the cave system again; it already covers about one-half. The cave refugium will then be reestablished, with no perceptible change since it was last a refugium.

Respites in the Neoglaciation

It was noted previously (and see chapter 1 and fig. 1.6) that there have been respites in the neoglaciation because the little ice age cycle went some way toward masking the general cooling trend. The most recent respite was the Little Climatic Optimum, which peaked about 1.8k B.P. It was the last warm spell before the current one, which began somewhat more than 100 years ago; between them came the Little Ice Age. The unusual cold of the Little Ice Age is part of recent human history; it earned the period its name, which is always written with capital initials; the phrase *little ice age cycle* is more recent and was coined to describe the climatic fluctuations of which the Little Ice Age is a part.

The Little Climatic Optimum left evidence of ecological effects in a number of places. One such place is Hell's Kitchen Lake (see fig. 14.3) in northern Wisconsin, where sediment cores yielded tree pollen, birch seeds, and charcoal fragments remaining from forest fires.[21] The presence of birch seeds makes it possible to know which species of birch were present; the pollens of the different birches are indistinguishable. Sediment layers dating from about 1.7k to 1.2k B.P. contain comparatively high proportions of paper birch seeds, oak and aspen pollen, and charcoal fragments. In contrast, older and younger sediments contain more seeds of yellow birch than of paper birch and fewer charcoal fragments. The findings imply that for about 500 years conditions were drier and warmer and forest fires more frequent than they were before or have been since.

Observations of a similar kind[22] made at Marion Lake in lower Michigan, about 350 kilometers to the east-southeast, also indicate a climatic optimum preceding the Little Ice Age. But in this case the timing is different: the warm period lasted for only 200 years and started much later, at about 1k B.P. Similarly (perhaps "dissimilarly" would be a better word) at Clearwater Bog, near The Pas, Manitoba,[23] 1,200 kilometers to the northwest of Hell's Kitchen Lake, the warm dry period seems to have lasted from 1.2k to 0.9k B.P. Assuming the evidence to have been correctly interpreted in each case, what seems to have been the Little Climatic Optimum happened at different times in different places, but there was *not* a geographical trend; the warmth came first to the middle site of the three, Hell's Kitchen Lake.

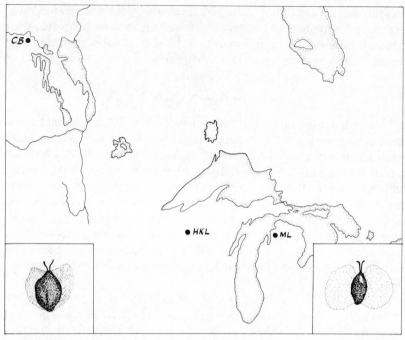

FIGURE 14.3: Locations of: Clearwater Bog, Manitoba, CB; Hell's Kitchen Lake, Wisconsin, HKL; Marion Lake, Michigan, ML. Inset: Left, seed of paper birch. Right, seed of yellow birch.

Examples like these could be multiplied. A possible explanation of the discrepancies in timing is that ecological inertia delayed the vegetation responses at Marion Lake and Clearwater Bog. There was probably no delay at Hell's Kitchen Lake, where abundant charcoal fragments (suggesting frequent forest fires, hence a warm, dry climate) are at the same level in the sediment as the pollen and seeds of drought-tolerant trees.

Pollen cores from many places show no evidence at all for any warm period after the onset of the neoglaciation. Examples come from near Vancouver, British Columbia,[24] near Edmonton, Alberta,[25] and near the delta of the Mackenzie River, Northwest Territories.[26] This does not mean that there was no climatic optimum at these places, only that it had no effect (or, strictly, no effect observable at the present day) on the vegetation.

This is not surprising. Recall (chapter 4) that climatic fluctuations in the last millennia of the Pleistocene, as the ice sheets were disappearing, seem not to have influenced the development of vegetation, which proceeded smoothly, with no reversals. Which raises the following question: assuming the little ice age cycle has been going on for at least 20,000 years, why did it affect vegetation in the recent past (at least in some places) and not in the more distant past? A possible answer is that vegetation on well-developed organic soils is more sensitive to climatic change than vegetation on the poorly developed, infertile soils on fresh glacial outwash.[27]

There appears to have been a Little Climatic Optimum at sea as well as on land. Shells of the bay scallop dating from 1.8k B.P. and 1.4k B.P. have been found[28] on the shores of Sable Island (see fig. 13.2), but the waters are too cold for them there now and too cold also for the eastern oysters that lived there in the hypsithermal (see chapter 13). The temperature of the inshore waters these species lived in were probably not the same as contemporary ocean temperatures. Sea level was rising; therefore, warm, shallow bays and lagoons probably formed from time to time and then disappeared. All the same, both the hypsithermal and the Little Climatic Optimum seem to have left their mark on Sable Island molluscs.

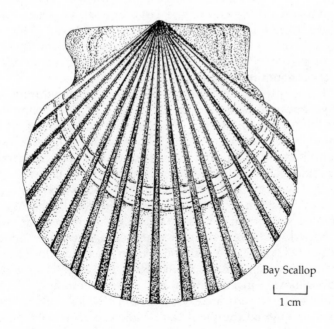

Bay Scallop

1 cm

The Little Ice Age

The most distinctive climatic interval of the recent past is the Little Ice Age, which lasted from about 1350 to 1870 A.D. or from 0.64k to 0.12k B.P. (With dates so recent the B.P. dating system has obvious drawbacks.) Glaciers and ice caps expanded, the ranges of tree species changed, and human beings adapted to the sudden cold that followed the Little Climatic Optimum.

Many Rocky Mountain glaciers advanced between one and two kilometers, injuring trees in the process. Many of the trees scarred by ice are still alive, and the date at which the damage happened can be discovered by counting the number of annual rings that have formed outside the scar. Other trees were pushed by the ice and tilted; because the annual rings of a leaning conifer are asymmetrical, being wider on the lower side, the number of years since a living tree was tilted can be learned by counting the wide rings outside the narrow ones. Trees were also overridden and killed; broken stumps of some of them are now exposed where glacier snouts have receded again, in the climatic warming that began a little over 100 years ago.[29] No doubt there are many more still under the ice.

Geomorphological signs of the ice advance, too, are plainly visible in the Rockies. The end moraines deposited by the glaciers when they retreated at the end of the Little Ice Age are still so new that they are bare of vegetation; they form a typical part of the scene in many a glacial valley, just a few hundred meters downslope from the present glacier snout. Older moraines, dating from the hypsithermal, are now covered with plants.

The ice caps in the arctic islands also expanded during the Little Ice Age but in a manner quite different from that of mountain glaciers. The mountain glaciers were flowing and sheared off trees in the process. The arctic ice caps grew passively, when the warmth of summer became insufficient to melt snowbanks at their margins; the snow accumulated year after year, turning to ice as it was compressed under its own weight. The new ice did not move, and the vegetation beneath it was merely flattened. Now, where Little Ice Age ice has recently receded, mats of dead vegetation, squashed flat but otherwise undamaged, are sometimes uncovered. A whole community of dead but well-preserved plants has been uncovered in Ellesmere Island (see fig. 13.2). The species were recognizable; they are the same as plants now growing nearby and include four-angled mountain heather, white dryad, arctic poppy, arctic willow, and purple saxifrage. Although all the plants are dead, and their seeds nonviable, the leaves of the mountain heather still contain one-fourth as much chlorophyll as is found in living leaves at the height of summer.[30]

The only perennial ice yielded by the Little Ice Age was that which was added to preexisting glaciers and ice caps. Elsewhere, winters were colder, summers cooler, and precipitation greater. Ecosystems of all kinds were affected, including those with human beings as members.

The coming of the neoglaciation had probably benefited the human population of the continent to begin with: populations had increased in size, technology had advanced, and hunters and gatherers had added gardening to their methods of obtaining food. At first only local plants had been grown. Later corn was brought north, and its cultivation became widespread. The intense cold of the Little Ice Age, however, caused a setback; the northern limit of what was then the corn belt was forced southward. Archaeological findings[31] show that, at a site near Winnipeg, corn growing was abandoned about 400 years ago, presumably because the growing season had become too short for it; the inhabitants of the site (ancestors of the modern Cree) went back to their traditional diet of animals and plants native to the region. This is the only archaeological indication of the colder climate. As to how daily life was affected, we can only speculate.

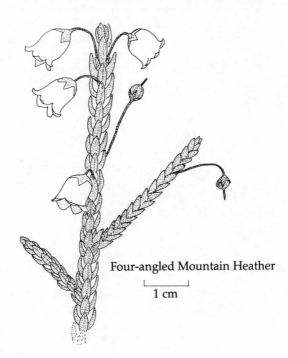

Four-angled Mountain Heather

1 cm

From the time the European invaders of North America established themselves and began keeping records, the bitter winters of the Little Ice Age become part of written history.

From that point also, the natural history of northern North America began to deviate from its "natural" course. The continent was no longer isolated. The foreign invaders multiplied rapidly, destroying native ecosystems at an ever increasing rate. In time, the byproducts of technology began to poison earth, water, and air and have now begun to influence the climate. The measured responses of biosphere to climate, and of climate to astronomical controls have, for the foreseeable future, come to an end. And the story told in this book comes, at least temporarily, to a close.

Epilogue

What of the future? The human population explosion with its accompanying technological explosion is disrupting the orderly development of the world's biosphere in a variety of ways. At the time of writing (New Year's, 1990), the greenhouse effect appears to be the paramount disruption. Predictions vary as to the magnitude of the climatic change we should expect from this cause; some climatologists foresee an increase in annual mean temperatures throughout the world of as much as five degrees Celsius in fifty years.

Such a change, if it comes, will be unlike anything this continent has undergone since the disappearance of Wisconsin ice. A change of such magnitude in so short a time may be wholly unprecedented. (The rate at which the temperature rose during the hypsithermal, for example, was probably not more than one degree Celsius per 500 years; now it could be as high as one degree Celsius per decade.) Attempts to predict the future therefore seem futile; we have no past experience to build on. Perhaps increasing cloudiness will compensate for the rising concentrations of greenhouse gases in the atmosphere and yield a different set of consequences. In any case, the whole world is affected, not merely this continent.

Although anthropocentrism is usually unscientific, it does seem safe to say that "now" is a special time, a transition period; and that we (*Homo sapiens*) are a special species, the unwitting cause of the transition. One thing seems certain: the world will eventually resume its orderly development after our species goes extinct. Questions about what will happen before that, before (as one might say) natural order is restored, are unanswerable.

Perhaps, fifty million years hence, members of some new species of animal, interested in paleoecology, will examine the geological record. They will note the transition that happened, from their point of view, fifty million years ago. Will they regard it as major or minor? Will it rate as the end of the Holocene Epoch or even as the end of the Quaternary Period? Or will it be no more than a minor episode early in the great Quaternary Glacial Age, during which some species (including *Homo sapiens*, perhaps?) went extinct and others shifted their ranges?

We shall never know.

Appendix 1

Common (English) names and scientific (Latin) names of all species mentioned in this book.

alder	*Alnus* spp.
red	*A. rubra*
arborvitae ("cedar")	*Thuja* spp.
arnica, heart-leaved	*Arnica cordifolia*
arrowhead	*Sagittaria* spp.
broad-leaved	*S. latifolia*
ash	*Fraxinus* spp.
black	*F. nigra*
white	*F. americana*
aspen	*Populus tremuloides*
ass	*Equus* spp.
badger	*Taxidea taxus*
basswood	*Tilia americana*
bayberry	*Myrica pennsylvanica*
bear	*Ursus* spp.
giant short-faced	*Arctodus simus*
grizzly	*Ursus arctos*
beardtongue, blue	*Penstemon nitidus*
bear grass	*Xerophyllum tenax*
beaver	*Castor canadensis*
giant	*Castoroides ohioensis*
beech	*Fagus grandifolia*
beetle, rove	*Homaeotarsus lecontei*
beluga	*Delphinapterus leucas*
betony, wood	*Pedicularis bracteosa*

bighorn sheep	*Ovis canadensis*
birch	*Betula* spp.
dwarf	*B. nana*
paper	*B. papyrifera*
shrub	*B. glandulosa*
yellow	*B. alleghaniensis*
bison	*Bison* spp.
blackbird, red-winged	*Agelaius phoeniceus*
blackfish, Alaska	*Dallia pectoralis*
blazing star	*Liatris punctata*
blueberry, bog	*Vaccinium uliginosum*
bluethroat	*Luscinia svecica*
bobcat	*Lynx rufus*
bracken	*Pteridium aquîlinum*
buckbean	*Menyanthes trifoliata*
buffalo berry	*Shepherdia canadensis*
thorny	*S. argentea*
bulrush, seacoast	*Scirpus maritimus*
bunchberry	*Cornus canadensis*
burbot	*Lota lota*
bur reed	*Sparganium* spp.
butterfly, Melissa arctic	*Oeneis melissa*
butterfly weed	*Gaura coccinea*
cactus, cushion	*Coryphantha vivipara*
calypso orchid	*Calypso bulbosa*
camas, blue	*Camassia quamash*
camel, western	*Camelops hesternus*
campion, moss	*Silene acaulis*
caribou	*Rangifer tarandus*
catbrier	*Smilax rotundifolia*
cat, scimitar	*Homotherium serum*
cattail	*Typha latifolia*
cedar, yellow	*Chamaecyparis nootkatensis*
chestnut	*Castanea* spp.
cheetah, American	*Miracinonyx trumani*
chickadee, gray-headed	*Parus cinctus*
chipmunk, eastern	*Tamias striatus*
clam, soft-shell	*Mya arenaria*
cloudberry	*Rubus chamaemorus*
cod	*Gadus morhua*
cone flower	*Ratibida columnifera*
coontail	*Ceratophyllum demersum*

coot	*Fulica americana*
corydalis, pale	*Corydalis pauciflora*
cotton grass	*Eriophorum* spp.
cougar	*Felis concolor*
cottonwood	*Populus* spp.
cranberry	*Vaccinium oxycoccos*
crowberry	*Empetrum nigrum*
currant, sticky	*Ribes viscosissimum*
deer	*Odocoileus* spp.
deer (extinct)	*Navahoceros* spp.
	Sangamona spp.
diapensia, Lapland	*Diapensia lapponica*
dock	*Rumex* spp.
dog, domestic	*Canis familiaris*
Douglas-fir	*Pseudotsuga menziesii*
dryad	*Dryas* spp.
white	*D. intergrifolia*
yellow	*D. drummondii*
duck	
ruddy	*Oxyura jamaicensis*
wood	*Aix sponsa*
elk; *see* wapiti	
elm	*Ulmus* spp.
white	*U. americana*
ermine	*Mustela erminea*
fairy candelabra	*Androsace septentrionalis*
fern, holly	*Polystichum lonchitis*
fir	*Abies* spp.
alpine	*A. lasiocarpa*
balsam	*A. balsamea*
fireweed	*Epilobium angustifolium*
mountain	*E. latifolium*
fisher	*Martes pennanti*
flax, blue	*Linum perenne*
flicker	*Colaptes auratus*
fox	
gray	*Urocyon cinereoargenteus*
red	*Vulpes vulpes*
swift	*V. velox*
fritillary	*Fritillaria* spp.

gentian, bog	*Gentiana calycosa*
glasswort	*Salicornia* spp.
goat, mountain	*Oreamnos americanus*
golden-bean	*Thermopsis rhombifolia*
goldenrod	*Solidago* spp.
mountain	*S. multiradiata*
goldeye	*Hiodon alosoides*
grass	
beach	*Ammophila breviligulata*
cord	*Spartina alterniflora*
grama	*Bouteloua gracilis*
grayling, arctic	*Thymallus arcticus*
grebe	
horned	*Podiceps auritus*
western	*Aechmophorus occidentalis*
ground squirrel	
arctic	*Spermophilus parryii*
Richardson's	*S. richardsonii*
thirteen-lined	*S. tridecemlineata*
hare	*Lepus* spp.
harebell, mountain	*Campanula lasiocarpa*
hare, snowshoe	*Lepus americanus*
heather	
four-angled mountain	*Cassiope tetragona*
yellow	*Phyllodoce glanduliflora*
hemlock	*Tsuga* spp.
eastern	*T. canadensis*
western	*T. heterophylla*
hemlock looper	*Lamdina fiscellaria*
hickory	*Carya* spp
honeysuckle, Utah	*Lonicera utahensis*
horse	*Equus* spp.
human beings	*Homo sapiens*
jackrabbit	*Lepus townsendii*
junco	*Junco hyemalis*
juniper	*Juniperus* spp.
kangaroo rat, Ord's	*Dipodomys ordii*
kelp	*Laminaria* spp.
killifish, banded	*Fundulus diaphanus*

Labrador tea	*Ledum palustre*
lamb's quarters	*Chenopodium album*
larch (tamarack)	*Larix laricina*
laurel, swamp	*Kalmia polifolia*
leatherleaf	*Chamaedaphne calyculata*
lemming	*Dicrostonyx* spp.
Ungava	*D. hudsonius*
brown	*Lemmus* spp.
lion, American	*Panther leo atrox*
llama	*Paleolama* spp.
lousewort, woolly	*Pedicularis lanata*
lovage	*Ligusticum* spp.
lupine, arctic	*Lupinus arcticus*
mallow, apricot	*Sphaeralcea coccinea*
mammoth, woolly	*Mammuthus primigenius*
maple	*Acer* spp.
sugar	*A. saccharum*
marmot	*Marmota* spp.
hoary	*M. caligata*
marsh marigold, mountain	*Caltha leptosepala*
marten	*Martes americana*
mastodon	*Mammut americanum*
milfoil, water	*Myriophyllum* spp.
mink	*Mustela vison*
minnow, brassy	*Hybognathus hankinsoni*
monkshood, mountain	*Aconitum delphinifolium*
moose	*Alces alces*
moss	
feather	*Hylocomium splendens*
	Pleurozium schreberi
	Ptilium crista-castrensis
peat	*Sphagnum* spp.
mountain-box	*Pachistima myrsinites*
mouse, western harvest	*Reithrodontomys megalotis*
muskellunge	*Esox masquinongy*
muskox	
tundra	*Ovibos moschatus*
woodland	*Symbos cavifrons*
muskrat	*Ondatra zibethicus*
mussel, blue	*Mytilus edulis*

naiad — *Najas flexilis*
narwhal — *Monodon monoceros*
nematode, swimbladder — *Cystidicola farionis*

oak — *Quercus* spp.
  burr — *Q. macrocarpa*
orache — *Atriplex* spp.
otter
  river — *Lontra canadensis*
  sea — *Enhydra lutris*
owl, short-eared — *Asio flammeus*
oyster, eastern — *Crassostrea virginica*
oysterleaf — *Mertensia maritima*

paintbrush, Indian — *Castilleja* spp.
partridge foot — *Luetkea pectinata*
pea, beach — *Lathyrus* spp.
phlox, moss — *Phlox hoodii*
pigweed — *Chenopodium* spp.
pike, northern — *Esox lucius*
pine — *Pinus* spp.
  jack — *P. banksiana*
  limber — *P. flexilis*
  lodgepole — *P. contorta*
  red — *P. resinosa*
  shore — *P. contorta contorta*
  western white — *P. monticola*
  eastern white — *P. strobus*
  whitebark — *P. albicaulis*
pipsissewa — *Chimaphila umbellata*
plantain, common — *Plantago major*
pondweed — *Potamogeton* spp.
poplar — *Populus* spp.
  balsam — *P. balsamifera*
poppy, arctic — *Papaver lapponicum*
porpoise, harbor — *Phocoena phocoena*
prairie clover, white — *Petalostemum candidum*
prairie smoke — *Geum triflorum*
prickly pear — *Opuntia polyacantha*
pronghorn — *Antilocapra americana*

ragweed — *Ambrosia artemisifolia*
ragwort — *Senecio* spp.

reindeer moss	*Cladonia* spp.
rocket, sea	*Cakile edentula*
rosemary, bog	*Andromeda polifolia*
rush	*Juncus* spp.
Russian thistle	*Salsola kali*
sabertooth	*Smilodon fatalis*
sage (sagebrush)	*Artemisia* spp.
pasture	*A. frigida*
salamander, tiger	*Ambystoma tigrina*
salmon, Pacific	*Oncorhynchus* spp.
salt hay	*Spartina patens*
sandwort	*Arenaria* spp.
saxicave, arctic	*Hiatella arctica*
saxifrage	
golden	*Saxifraga aizoides*
purple	*S. oppositifolia*
scallop	
bay	*Aquipecten irradians*
Iceland	*Chlamys islandica*
scimitar cat	*Homotherium serum*
sculpin	
four-horned	*Myxocephalus quadricornis*
prickly	*Cottus asper*
seacow, Steller's	*Hydromalis gigas*
seal	
bearded	*Erignathus barbatus*
fur	*Callorhinus ursinus*
harp	*Phoca groenlandica*
ringed	*P. hispida*
sea lion, northern	*Eumetropias jubata*
sedge	*Carex* spp.
large-headed	*C. macrocephala*
sheep, Dall's	*Ovis dalli*
shiner, blacknose	*Notropis heterolepis*
shrew	
smoky	*Sorex fumeus*
dwarf	*S. nanus*
northern water	*S. palustris*
shrimp	
brine	*Artemia salina*
opossum	*Mysis relicta*

shrike
  loggerhead   *Lanius ludovicianus*
  northern   *L. excubitor*
shrub-ox   *Euceratherium collinum*
sibbaldia   *Sibbaldia procumbens*
silverberry (wolf willow)   *Eleagnus commutatus*
skunk cabbage   *Symplocarpos foetidus*
sloth, Shasta ground   *Nothrotheriops shastensis*
snail
  stagnant pond   *Lymnaea stagnalis*
  tow-ridged ramshorn   *Helisoma anceps*
  tadpole   *Physa gyrina*
snake, garter   *Thamnopsis sirtalis*
sorrel   *Rumex* spp.
  mountain   *Oxyria digyna*
spatterdock   *Nuphar polysepalum*
spikemoss   *Selaginella* spp.
spleenwort, green   *Asplenium viride*
spring beauty   *Claytonia* spp.
  western   *C. lanceolata*
spruce   *Picea* spp.
  black   *P. mariana*
  Engelmann   *P. engelmannii*
  Sitka   *P. sitchensis*
  white   *P. glauca*
spurge, seaside   *Euphorbia polygonifolia*
squirrel, ground; *see* ground
    squirrel
stag-moose   *Cervalces scottii*
stickleback
  brook   *Culaea inconstans*
  three-spined   *Gasterosteus aculeatus*
stonewort   *Chara* spp.
strawberry
  coastal   *Fragaria chiloensis*
  wild   *F. virginiana*
sucker, mountain   *Catostomus platyrhynchus*
sundew   *Drosera* spp.
swallow, cliff   *Hirundo pyrrhonota*
sweet gale   *Myrica gale*

tamarack; *see* larch
thimbleberry — *Rubus parviflorus*
toad, plains spadefoot — *Scaphiopus bombifrons*
tomcod — *Microgadus tomcod*
trout, lake — *Salvelinus namaycush*
twinflower — *Linnaea borealis*

violet — *Viola* spp.
vole — *Microtus* spp.
  Gapper's red-backed — *Clethryonomis gapperi*
  heather — *Phenacomys intermedius*

wagtail
  white — *Motacilla alba*
  yellow — *M. flava* .
walrus — *Odobenus rosmarus*
wapiti (elk) — *Cervus elephas*
warbler, yellow-rumped — *Dendroica coronata*
waterlily, yellow; see spatter-
  dock
waxwing
  Bohemian — *Bombycilla garrulus*
  cedar — *B. cedrorum*
weasel, least — *Mustela nivalis*
whale
  bowhead — *Balaena mysticetus*
  finback — *Balaenoptera physalis*
  humpback — *Megaptera novaeangliae*
whelk, Neptune — *Neptunea despecta*
whitefish, lake — *Coregonis clupeaformis*
willow — *Salix* spp.
  arctic — *S. arctica*
withe-rod — *Viburnum* spp.
wolf
  dire — *Canis dirus*
  timber — *C. lupus*
wolverine — *Gulo gulo*
woodpecker
  black-backed three-toed — *Picoides arcticus*
  northern three-toed — *P. tridactylus*
wormwood; *see* sage

Appendix 2

Scientific (Latin) names and common (English) names of all species
mentioned in this book

Abies spp.	fir
A. balsamea	balsam
A. lasiocarpa	alpine
Abronia latifolia	verbena, yellow sand
Acer spp.	maple
A. saccharum	sugar
Aconitum delphinifolium	monkshood, mountain
Aechmophorus occidentalis	grebe, western
Aegolius acadicus	owl, saw-whet
A. funereus	boreal
Agelaius phoeniceus	blackbird, redwinged
Aix sponsa	duck, wood
Alces alces	moose
Alnus spp.	alder
Ambrosia artemisifolia	ragweed
Ambystoma tigrinum	salamander, tiger
Ammophila spp.	beach grass
Andromeda polifolia	rosemary, bog
Androsace septentrionalis	fairy candelabra
Antilocapra americana	pronghorn
Aquipecten irradians	scallop, bay
Arctodus simus	bear, giant short-faced
Arenaria spp.	sandwort
Arnica cordifolia	arnica, heart-leaved
Artemia salina	shrimp, brine

Artemisia spp.	sage (sagebrush)
A. frigida	pasture
Asio flammeus	owl, short-eared
Asplenium viride	spleenwort, green
Atriplex spp.	orache
Balaena mysticetus	whale, bowhead
Balaenopteris physalis	finback
Betula spp.	birch
B. alleghaniensis	yellow
B. glandulosa	shrub
B. nana	dwarf
B. papyrifera	paper
Bison spp.	bison
Bombycilla cedrorum	waxwing, cedar
B. garrulus	Bohemian
Bouteloua gracilis	grass, grama
Cakile edentula	rocket, sea
Callorhinus ursinus	seal, fur
Caltha leptosepala	marsh marigold, mountain
Calypso bulbosa	calypso orchid
Camassia quamash	camas, blue
Camelops hesternus	camel, western
Campanula lasiocarpa	harebell, mountain
Canis dirus	wolf, dire
C. lupus	timber
C. familiaris	dog, domestic
Capromerix sp.	pronghorn (extinct)
Carex spp.	sedge
C. macrocephala	large-headed
Carya spp.	hickory
Cassiope tetragona	mountain heather, four-angled
Castanea spp.	chestnut, sweet
Castilleja spp.	paintbrush, Indian
Castor canadensis	beaver
Castoroides ohioensis	giant
Catostomus platyrhynchus	sucker, mountain
Ceratophyllum demersum	coontail
Cervalces scottii	stag-moose
Cervus elephas	wapiti
Chamaecyparis nootkatensis	cedar, yellow
Chamaedaphne calyculata	leatherleaf

Chara spp.	stonewort
Chenopodium spp.	pigweed
C. album	lamb's quarters
Chimaphila umbellata	pipsissewa
Chlamys islandica	scallop, Iceland
Cladonia spp.	reindeer moss
Claytonia spp.	spring beauty
Clethrionomys gapperi	vole, Gapper's red-backed
Colaptes auratus	flicker
Coregonus clupeaformis	whitefish, lake
Cornus canadensis	bunchberry
Corydalis pauciflora	corydalis, pale
Coryphantha vivipara	cactus, cushion
Cottus asper	sculpin, prickly
Crassostrea virginica	oyster, eastern
Culaea inconstans	stickleback, brook
Cystidicola farionis	nematode, swimbladder
Dallia pectoralis	blackfish, Alaska
Delphinapterus leucas	beluga
Dendroica coronata	warbler, yellow-rumped
Diapensia lapponica	diapensia, Lapland
Dicrostonyx spp.	lemming
D. hudsonius	Ungava
D. torquatus	collared
Dipodomys ordii	kangaroo rat, Ord's
Drosera spp.	sundew
Dryas spp.	dryad
D. drummondii	yellow
D. integrifolia	white
Eleagnus commutatus	silverberry, wolf willow
Empetrum nigrum	crowberry
Enhydra lutris	otter, sea
Epilobium spp.	fireweed
E. latifolium	mountain
Equus spp.	horse, ass
E. conversidens	Mexican
E. lambei	Lambe's
Erignathus barbatus	seal, bearded
Eriophorum spp.	cotton grass
Esox lucius	pike, northern
E. masquinongy	muskellunge

Eumetropias jubata	sea lion, northern
Euceratherium collinum	shrub-ox
Euphorbia polygonifolia	spurge, seaside
Fagus grandifolia	beech
Felis concolor	cougar
Fragaria chiloensis	strawberry, coastal
Fraxinus spp.	ash
F. americanus	white
Fritillaria spp.	fritillary
Fulica americana	coot
Gadus morhua	cod
Gasterosteus aculeatus	stickleback, three-spined
Gaura coccinea	butterfly weed
Gentiana calycosa	gentian, bog
Geum triflorum	prairie smoke
Gulo gulo	wolverine
Helisoma anceps	snail, two-ridged ramshorn
Hiatella arctica	saxicave, arctic
Hiodon alosoides	goldeye
Hirundo pyrrhonota	swallow, cliff
Homaeotarsus lecontei	beetle, rove
Homo sapiens	human beings
Homotherium serum	cat, scimitar
Hybognathus hankinsoni	minnow, brassy
Hydromalis gigas	seacow, Steller's
Hylocomium splendens	moss, feather
Junco hyemalis	junco
Juncus spp.	rush
Juniperus spp.	juniper
Kalmia polifolia	laurel, swamp
Lamdina fiscellaria	hemlock looper
Laminaria spp.	kelp
Lampsilis radiata	clam, lampsilis
Lanius excubitor	shrike, northern
L. ludovicianus	loggerhead
Larix laricina	larch or tamarack

Lathyrus spp.	beach-pea
Ledum palustre	Labrador tea
Lemmus spp.	lemming, brown
Lepus americanus	hare, snowshoe
L. townsendii	jackrabbit
Liatris punctatum	blazing star
Ligusticum spp.	lovage
Linnaea borealis	twinflower
Linum perenne	flax, blue
Lonicera utahensis	honeysuckle, Utah
Lontra canadensis	otter, river
Lota lota	burbot
Luscinia svecica	bluethroat, Alaska
Luetkea pectinata	partridge-foot
Lupinus arcticus	lupine, arctic
Lymnaea stagnalis	snail, stagnant pond
Lynx rufus	bobcat
Mammut americanum	mastodon
Mammuthus primigenius	mammoth, woolly
Marmota spp.	marmot
M. caligata	hoary
Martes americana	marten
M. pennanti	fisher
Megaptera novaeangliae	whale, humpback
Menyanthes trifoliata	buckbean
Mertensia maritima	oysterleaf
Microgadus tomcod	tomcod
Microtus spp.	vole
Miracinonyx trumani	cheetah, American
Monodon monoceros	narwhal
Motacilla alba	wagtail, white
M. flava	yellow
Mustela nivalis	weasel, least
M. vison	mink
Mya arenaria	clam, soft-shell
Myrica gale	sweet gale
M. pennsylvanica	bayberry
Myriophyllum spp.	milfoil, water
Mysis relicta	shrimp, opossum
Mytilus edulis	mussel, blue
Myxocephalus quadricornis	sculpin, four-horned

Najas flexilis	naiad
Navahoceros sp.	deer (extinct)
Neptunea despecta	whelk
Nothrotheriops shastensis	sloth, Shasta ground
Notropis heterolepis	shiner, blacknose
Nuphar polysepalum	spatterdock
Odobenus rosmarus	walrus
Odocoileus spp.	deer
Oeneis melissa	butterfly, Melissa arctic
Oncorhynchus spp.	salmon, Pacific
Ondatra zibethicus	muskrat
Opuntia polyacantha	prickly pear
Oreamnos americanus	goat, mountain
Ovibos moschatus	muskox, tundra
Ovis canadensis	sheep, bighorn
O. dalli	Dall's
Oxyria digyna	sorrel, mountain
Pachistima myrsinites	mountain-box
Paleolama spp.	llama
Panthera leo atrox	lion, American
Papaver lapponicum	poppy, arctic
Parus cinctus	chickadee, gray-headed
Pedicularis bracteosa	betony, wood
P. lanata	lousewort, woolly
Penstemon nitidus	beardtongue, blue
Petalostemum candidum	prairie clover, white
Phenacomys intermedius	heather vole
Phlox hoodii	phlox, moss
Phoca groenlandica	seal, harp
P. hispida	ringed
Phocoena phocoena	porpoise, harbour
Phyllodoce glanduliflora	heather, yellow
Physa gyrina	snail, tadpole
Picea spp.	spruce
P. engelmannii	Engelmann's
P. glauca	white
P. mariana	black
P. sitchensis	Sitka
Picoides arcticus	woodpecker, black-backed three-toed
P. tridactylus	northern three-toed

Pinus spp.	pine
P. *albicaulis*	whitebark
P. *banksiana*	jack
P. *contorta*	lodgepole
P. *contorta contorta*	shore
P. *flexilis*	limber
P. *monticola*	western white
P. *strobus*	eastern white
Plantago major	plantain, common
Pleurozium schreberi	moss, feather
Podiceps auritus	grebe, horned
Polemonium viscosum	skunkweed
Polystichum lonchitis	fern, holly
Populus spp.	poplar
P. *balsamifera*	balsam
P. *tremuloides*	aspen
Potamogeton spp.	pondweed
Pseudotsuga menziesii	Douglas-fir
Pteridium aquilinum	bracken
Ptilium crista-castrensis	moss, feather
Quercus spp.	oak
Q. *macrocarpa*	burr
Rangifer tarandus	caribou
Ratibida columnifera	cone flower
Reithrodontomys megalotis	harvest mouse, western
Ribes viscosissimum	currant, sticky
Rubus chamaemorus	cloudberry
R. *parviflorus*	thimbleberry
Rumex spp.	dock
Sagittaria latifolia	arrowhead, broad-leaved
Saiga tatarica	saiga antelope
Salicornia spp.	glasswort
Salix spp.	willow
S. *arctica*	arctic
Salsola kali	Russian thistle
Salvelinus namaycush	trout, lake
Sangamona sp.	deer (extinct)
Saxifraga aizoides	saxifrage, golden
S. *oppositifolia*	purple
Scaphiopus bombifrons	toad, plains spadefoot

Scirpus maritimus	bulrush, seacoast
Selaginella spp.	spikemoss
Senecio spp.	ragwort
Shepherdia argentea	buffalo berry, thorny
S. canadensis	buffalo berry
Sibbaldia procumbens	sibbaldia
Silene acaulis	campion, moss
Smilax rotundifolia	cat-brier
Smilodon fatalis	sabertooth
Solidago multiradiata	goldenrod, mountain
Sorex fumeus	shrew, smoky
S. nanus	dwarf
S. palustris	northern water
Sparganium spp.	bur reed
Spartina alterniflora	cord grass
S. patens	salt hay
Spermophilus parryi	ground squirrel, arctic
S. richardsonii	Richardson's
S. tridecemlineatus	thirteen-lined
Sphaeralcea coccinea	mallow, apricot
Sphagnum spp.	moss, peat
Stockoceros sp.	pronghorn (extinct)
Symbos cavifrons	muskox, woodland
Symplocarpos foetidus	skunk cabbage
Tamias striatus	chipmunk, eastern
Taxidea taxus	badger
Thamnopsis sirtalis	snake, garter
Thermopsis rhombifolia	golden-bean
Thuja spp.	arborvitae
Thymallus arcticus	grayling, arctic
Tilia americana	basswood
Tsuga spp.	hemlock
T. canadensis	eastern
T. heterophylla	western
Typha latifolia	cattail
Ulmus spp.	elm
U. americana	white
Urocyon cinereoargenteus	fox, gray
Ursus arctos	bear, grizzly

Vaccinium oxycoccos	cranberry
V. uliginosum	blueberry, bog
Viburnum spp.	withe-rod
Vulpes velox	fox, swift
V. vulpes	red
Xerophyllum tenax	bear grass

Notes

Chapter 1

1. J. Imbrie and K. P. Imbrie, 1979, *Ice ages: Solving the mystery* (Short Hills, N.J.: Enslow Publishers). See also J. Gribbin, 1982, *Future weather* (New York: Penguin Books).

2. Imbrie and Imbrie, *Ice ages.*

3. The maps in figures 1.2, 1.3, 1.4, and 1.5 are based on V. K. Prest, 1973, *Glacial retreat: Retreat of the last ice sheet* (National Atlas of Canada, Folio A:31–32, Surveys and Mapping Branch, Department of Energy, Mines and Resources, Ottawa) and on P. A. Mayewski, G. H. Denton and T. J. Hughes, 1981, The Last Wisconsin ice sheets in North America. In *The last great ice sheets*, ed. G. H. Denton and T. J. Hughes, 67–178 (New York: Wiley).

4. The figure is based on various sources, including J. Imbrie and J. Z. Imbrie, 1980, Modeling the climatic response to orbital variations, *Science* 207:943–53; G. H. Denton and W. Karlen, 1973, Holocene climatic variations—their pattern and possible cause, *Quaternary Research* 2:155–205; S. C. Porter and G. H. Denton, 1967, Chronology of neo-glaciation in the North American cordillera, *American Journal of Science* 265:177–210.

5. Imbrie and Imbrie, *Ice ages.*

6. H. Nichols, P. M. Kelly, and J. T. Andrews, 1978, Holocene palaeo-wind evidence from palynology in Baffin Island, *Nature* 273:140–2.

7. Gribbin, *Future weather.*

8. A. Neftel, H. Oeschger, J. Schwander, et al., 1982, Ice core sample measurements give atmospheric CO_2 content during the past 40,000 years, *Nature* 295:220–3.

9. M. B. Davis, 1984, Climatic instability, time lags, and community disequilibrium. In *Community ecology*, ed. J. Diamond and T. J. Case, 269–84 (New York: Harper and Row).

10. R. A. Watson and H. E. Wright, 1980, The end of the Pleistocene: a general critique of chronostratigraphic classification, *Boreas* 9:153–63.

11. H. D. Hedberg, 1976, *International stratigraphic guide: a guide to stratigraphic classification, terminology and procedure* (New York: Wiley).

12. Mayewski, Denton, and Hughes, Late Wisconsin Ice Sheets.

13. Porter and Denton, Chronology of Neo-glaciation.

14. E. L. Ladurie, 1971, *Times of feast, times of famine: a history of climate since the year 1000* (Translated by Barbara Bray). (New York: Doubleday).

15. A. S. Dyke and V. K. Prest, 1986, Late Wisconsinan and Holocene history of the Laurentide ice sheet, *Géographie Physique et Quaternaire* 41:237–64. Also, A. S. Dyke and V. K. Prest, 1987, Paleogeography of northern North America, 18 000–5 000 years ago, Geological Survey of Canada, Map 1703A, scale 1: 12 500 000.

16. Mayewski, Denton, and Hughes, Late Wisconsin Ice Sheet; Dyke and Prest, "Late Wisconsin Ice Sheets."

17. H. E. Wright, Jr. 1984, Sensitivity and response time of natural systems to climatic change in the late Quaternary, *Quaternary Science Reviews* 3:91–131.

18. Ibid.

19. G. H. Denton, 1974, Quaternary glaciations of the White River Valley, Alaska, with a regional synthesis for the St. Elias Mountains, Alaska and Yukon Territory, *Geological Society of America Bulletin* 85:871–92. Also, Mayewski, Denton, and Hughes, Late Wisconsin Ice Sheets.

20. J. G. Ogden, III, 1967, Radiocarbon and pollen evidence for a sudden change in climate in the Great Lakes region approximately 10,000 years ago. In *Quaternary paleoecology*, ed. E. J. Cushing and H. E. Wright, Jr., 117–27 (New Haven, Conn.: Yale University Press).

21. R. A. Bryson, D. A. Baerreis, and W. M. Wendland, 1970, The character of late-glacial and post-glacial climatic changes. In *Pleistocene and Recent environments of the Central Great Plains*, ed. W. Dort, Jr., J. K. Jones, Jr., 53–72 (Lawrence, Kan.: University of Kansas Press).

22. R. A. Bryson, 1985, On climatic analogs in paleoclimatic reconstruction, *Quaternary Research* 23:275–86.

23. CLIMAP Project Members, 1976, The surface of the ice-age earth, *Science* 191:1131–37.

24. D. M. Hopkins, 1982, Aspects of the paleogeography of Beringia during the Late Pleistocene. In *Paleoecology of Beringia*, ed. D. M. Hopkins, J. V. Matthews, Jr., C. E. Schweger, and S. B. Young, 3–28 (New York: Academic Press).

25. W. F. Ruddiman and A. McIntyre, 1981, Oceanic mechanisms for amplification of the 23,000 year ice-volume cycle, *Science* 212:617–27.

26. Ibid.

27. H. Nichols, 1976, Historical aspects of the northern Canadian treeline, *Arctic* 29:38–47.

28. R. F. Flint, 1971, *Glacial and Pleistocene geology* (New York: Wiley).

29. D. R. Grant, 1975, Recent coastal submergence of the Maritime Provinces. *Proceedings of the Nova Scotia Institute of Science*, Vol. 27, suppl. 3:83–102.

30. J. Clague, J. R. Harper, R. J. Hebda, and D. E. Howes, 1982, Late Quaternary sea level and crustal movements, coastal British Columbia, *Canadian Journal of Earth Sciences* 19:597–618.

31. R. J. Hebda and G. E. Rouse, 1979, Palynology of two Holocene cores from the Hesquiat Peninsula, Vancouver Island, British Columbia, *Syesis* 12:121–9.

32. Clague et al., Late Quaternary Sea Level.

33. L. Clayton and S. R. Moran, 1982, Chronology of Late Wisconsinan glaciation in Middle North America, *Quaternary Science Reviews* 1:55–82.

34. L. Clayton, 1967, Stagnant-glacier features of the Missouri Coteau in North Dakota. In *Glacial geology of the Missouri Coteau and adjacent areas*, ed. L. Clayton and T. F. Freers, 25–46, North Dakota Geological Survey: Miscellaneous Series 30.

35. C. E. Sloan, 1967, Groundwater movement as indicated by plants in the prairie pothole region. In Clayton and Freers, North Dakota Geological Survey.

36. E. A. Christiansen, 1979, The Wisconsinan deglaciation of southern Saskatchewan and adjacent areas, *Canadian Journal of Earth Sciences* 16:913–38.

37. C. B. Beaty, 1976, *The landscapes of southern Alberta: a regional geomorphology* (Lethbridge, Alberta: University of Lethbridge).

38. S. E. Stiegeler, ed., 1976, *A dictionary of earth sciences* (London: The Macmillan Press).

39. R. A. Bryson and W. M. Wendland, 1967, Tentative climatic patterns for some Late Glacial and Post-Glacial episodes in central North America. In *Life, land and water*, Proceedings of the 1966 Conference on Environmental Studies of the Glacial Lake Agassiz Region.

40. Ibid.

41. Mayewski, Denton, and Hughes, Late Wisconsin Ice Sheets.

42. Bryson, On climatic analogs.

43. R. J. Rogerson, 1983, Geological evolution. In *Biogeography and ecology of the island of Newfoundland*, ed. G. R. South, 5–35 (The Hague: Junk).

44. E. Dahl, 1946, On different types of unglaciated areas during the ice ages and their significance to phytogeography, *New Phytologist* 45:225–42.

45. Rogerson, Geological evolution.

46. Dahl, On different types.

Chapter 2

1. W. I. Illman, J. McLachlan, and T. Edelstein, 1970, Marine algae of the Champlain Sea episode near Ottawa, *Canadian Journal of Earth Sciences* 7:1583–5.

2. T. M. Cronin, 1977, Late-Wisconsin marine environments of the Champlain Valley (New York, Quebec), *Quaternary Research* 7:238–53.

3. L. D. Delorme, 1969, Ostracodes as Quaternary paleoecological indicators, *Canadian Journal of Earth Sciences* 6:1471–6.

4. I. R. Walker, 1987, Chironomidae (Diptera) in paleoecology, *Quaternary Science Reviews* 6:29–40.

5. I. J. Bassett, C. W. Crompton, and J. A. Parmelee, 1978, *An atlas of*

airborne pollen grains and common fungus spores of Canada, Canada Department of Agriculture, Research Branch, Ottawa. Monograph No. 18.

6. R. B. Davis and G. L. Jacobson, Jr., 1985, Late glacial and early Holocene landscapes in northern New England and adjacent areas of Canada, *Quaternary Research* 23:341–68.

7. T. Webb, III, 1981, The past 11,000 years of vegetational change in eastern North America, *BioScience* 31:501–6.

8. K. Faegri and J. Iversen, 1975, *Textbook of pollen analysis,* 3d ed. (New York: Hafner Press); Also, P. D. Moore and J. A. Webb, 1978, *An illustrated guide to pollen analysis* (London: Hodder and Stoughton).

9. The diagram is adapted and simplified from figure 8 of J. C. Ritchie, 1982, The modern and Late-Quaternary vegetation of the Doll Creek area, North Yukon, Canada, *New Phytologist* 90:563–603.

10. J. G. Ogden, III, 1986, An alternative to exotic spore or pollen addition in quantitative microfossil studies, *Canadian Journal of Earth Sciences* 23:102–6.

11. The diagram is adapted from figure 3 of J. C. Ritchie and L. C. Cwynar, The Late Quaternary vegetation of the north Yukon. In *Paleoecology of Beringia,* ed. D. M. Hopkins, J. V. Matthews, Jr., C. E. Schweger, and S. B. Young, (New York: Academic Press); and figure 9 in Ritchie, Modern and Late-Quaternary vegetation.

12. R. S. Bradley, 1985, *Quaternary paleoclimatology* (Boston: Allen and Unwin).

13. R. Burleigh, 1984, New World colonized in Holocene, *Nature* 312:399.

14. Bradley, *Quaternary paleoclimatology.*

15. Ibid.

Chapter 3

1. R. N. Mack, 1971, Pollen size variation in some western North American pines as related to fossil pollen identification, *Northwest Science* 45:257–69.

2. J. W. Ives, 1977, Pollen separation of three North American birches, *Arctic and Alpine Research* 9:73–80.

3. J. C. Ritchie and L. C. Cwynar, 1982, The Late Quaternary vegetation of the north Yukon. In *Paleoecology of Beringia,* ed. D. M. Hopkins, J. V. Matthews, Jr., C. E. Schweger, and S. B. Young, 113–26 (New York: Academic Press).

4. R. J. Mott and L. E. Jackson, Jr., 1982, An 18,000 year palynological record from the southern Alberta segment of the classical Wisconsin "ice-free corridor," *Canadian Journal of Earth Sciences* 19:504–13.

5. D. M. Hopkins, P. A. Smith, and J. V. Matthews, Jr., 1981, Dated wood from Alaska and the Yukon: Implications for forest refugia in Beringia, *Quaternary Research* 15:217–49.

6. A. G. Sangster and H. M. Dale, 1964, Pollen grain preservation of

underrepresented species in fossil spectra, *Canadian Journal of Botany* 42:437–49.

7. J. Terasmae, 1973, Notes on late Wisconsin and early Holocene history of vegetation in Canada, *Arctic and Alpine Research* 5:201–22.

8. The map is adapted from figure 3.3b in E. J. Crossman and D. E. McAllister, 1986, Zoogeography of freshwater fishes of the Hudson Bay drainage, Ungava Bay and the Arctic Archipelago. In *The zoogeography of North American freshwater fishes*, ed. C. H. Hocutt and E. O. Wiley, 17–104 (New York: Wiley-Interscience).

9. Crossman and McAllister, Zoogeography of freshwater fishes.

10. G. A. Black, 1983, *Cystidicola farionis* (Nematoda) as an indicator of lake trout (*Salvelinus namaycush*) of Bering ancestry, *Canadian Journal of Fisheries and Aquatic Sciences* 40:2034–40.

11. The map is adapted from figure 2 in Black, *Cystidicola farionis*.

12. C. S. Robbins, B. Bruun, and H. S. Zim, 1966, *Birds of North America* (New York: Golden Press).

13. R. M. Mengel, 1970, The North American Central Plains as an isolating agent in bird speciation. In *Pleistocene and Recent environments of the Central Great Plains*, ed. W. Dort and J. K. Jones, 279–340 (Lawrence, Kan., University of Kansas Press).

14. H. E. Wright, Jr., 1970, Vegetational history of the Central Plains. In Dort and Jones, *Pleistocene and Recent environments*, 157–72.

15. Mengel, North American Central Plains.

16. A. E. Porsild, 1958, Geographical distribution of some elements in the flora of Canada, *Geographical Bulletin* 11:57–77.

17. E. Hultén, 1968, *Flora of Alaska* (Stanford, Calif.: Stanford University Press).

Chapter 4

1. G. W. Argus and M. B. Davis, 1962, Macrofossils from a late-glacial deposit at Cambridge, Massachusetts, *American Midland Naturalist* 67:106–17.

2. T. L. Péwé, 1983, The periglacial environment in North America during Wisconsin time. In *Late Quaternary environments in the United States, vol. 1, The Late Pleistocene*, ed. S. C. Porter, 157–89 (Minneapolis: University of Minnesota Press).

3. W. A. Watts, 1983, Vegetational history of the eastern United States 25,000 to 10,000 years ago. In Porter, *Late-Quaternary environments*, 294–310.

4. C. R. Harington and A. C. Ashworth, 1986, A mammoth (*Mammuthus primigenius*) tooth from late Wisconsin deposits near Embden, North Dakota, and comments on the distribution of woolly mammoths south of the Wisconsin ice sheets, *Canadian Journal of Earth Sciences* 23:909–18.

5. Harington and Ashworth, A mammoth tooth. Also, L. D. Agenbroad, 1984, New World mammoth distribution. In *Quaternary extinctions: a prehistoric revolution*, ed. P. S. Martin and R. G. Klein, 90–108 (Tucson: University of Arizona Press).

6. Harington and Ashworth, A mammoth tooth. Also, B. Kurtén and E. Anderson, 1980, *Pleistocene mammals of North America* (New York: Columbia University Press).

7. P. A. Delcourt and H. R. Delcourt, 1981, Vegetation maps for eastern North America: 40,000 yr B.P. to the present. In *Geobotany II*, ed. R. C. Romans, 123–65 (New York: Plenum Press).

8. A. L. Washburn, 1973, *Periglacial processes and environments* (London: Edward Arnold).

9. D. B. Lawrence, 1958, Glaciers and vegetation in southeastern Alaska, *American Scientist* 46:89–122. Also, D. B. Lawrence, 1967, The role of *Dryas drummondii* in vegetation development following ice recession at Glacier Bay, Alaska, with special reference to its nitrogen-fixation by root nodules, *Journal of Ecology* 55:793–813.

10. L. B. Brubaker, 1975, Postglacial forest patterns associated with till and outwash in northcentral Upper Michigan, *Quaternary Research* 5:499–527.

11. M. B. Davis, 1981, Quaternary history and the stability of plant communities. In *Forest succession: Concepts and application*, ed. D. C. West, H. H. Shugart, and D. B. Botkin, 132–53 (New York: Springer-Verlag).

12. J. C. Ritchie and G. M. MacDonald, 1986, The patterns of postglacial spread of white spruce, *Journal of Biogeography* 13:527–40.

13. G. M. MacDonald and L. C. Cwynar, 1985, A fossil pollen based reconstruction of the late Quaternary history of lodgepole pine (*Pinus contorta* ssp. *latifolia*) in the western interior of Canada, *Canadian Journal of Forest Research* 15:1039–44.

14. L. C. Cwynar and G. M. MacDonald, 1987, Geographical variation of lodgepole pine in relation to population history, *American Naturalist* 129:463–9.

15. J. Terasmae, 1977, Postglacial history of Canadian muskeg. In *Muskeg and the northern environment in Canada*, ed. N. W. Radforth and C. W. Brawner, 9–30, (Toronto: University of Toronto Press).

16. D. R. Engstrom and B. C. S. Hansen, 1984, Postglacial vegetational change and soil development in southeastern Labrador as inferred from pollen and chemical stratigraphy, *Canadian Journal of Botany* 63:543–61.

17. H. F. Lamb, 1980, Late Quaternary vegetational history of southwestern Labrador *Arctic and Alpine Research* 12:117–35.

18. M. B. Davis, 1969, Climatic changes in southern Connecticut recorded by pollen deposition at Rogers Lake, *Ecology* 50:409–22.

19. J. C. Ritchie, 1984, *Past and present vegetation of the far northwest of Canada* (Toronto: University of Toronto Press).

20. E. Gorham, 1957, The development of peatlands, *Quarterly Review of Biology* 32:145–66.

21. D. M. Mickelson, L. Clayton, D. S. Fullerton, and H. W. Borns, Jr., 1983, The late Wisconsin glacial record of the Laurentide Ice Sheet in the United States. In *Late Quaternary environments of the United States, vol. 1, The Late Pleistocene*, ed. H. E. Wright, Jr., 3–37 (Minneapolis: University of Minnesota Press).

22. H. E. Wright, Jr., 1976, Ice retreat and revegetation of the western

Great Lakes area. In *Quaternary Stratigraphy of North America*, ed. W. C. Mahaney, 119–32 (Stroudsberg, Pa.: Dowden, Hutchinson and Ross Inc.).

23. M. B. Davis, 1984, Climatic instability, time lags, and community disequilibrium. In *Community ecology*, ed. J. Diamond and T. J. Case, 269–84 (New York: Harper and Row).

24. H. E. Wright, Jr., 1984, Sensitivity and response time of natural systems to climatic change in the late Quaternary, *Quaternary Science Reviews* 3:91–131.

25. H. E. Wright, Jr., 1976. The dynamic nature of Holocene vegetation. A problem in paleoclimatology, biogeography and stratigraphic nomenclature, *Quaternary Research* 6:581–96.

26. Wright, Jr., Sensitivity and response time; see figures 1 to 8 and 11. Also L. B. Brubaker and E. R. Cook, 1983, Tree-ring studies of Holocene environments. In *Late Quaternary environments of the United States, vol. 2, The Late Holocene*, ed. H. E. Wright, Jr., 222–35 (Minneapolis: University of Minnesota Press).

27. W. D. Billings and H. A. Mooney, 1968, The ecology of arctic and alpine plants, *Biological Reviews* 43:481–530.

Chapter 5

1. H. E. Wright, Jr., 1981. Vegetation east of the Rocky Mountains 18,000 years ago, *Quaternary Research* 15:113–25.

2. H. E. Wright, Jr., 1970, Vegetational history of the Central Plains. In *Pleistocene and Recent environments of the central Great Plains*, ed. W. Dort, Jr., J. K. Jones, Jr., 158–72 (Lawrence, Kan.: University of Kansas Press).

3. Wright, Jr., Vegetation east of the Rocky Mountains.

4. P. A. Delcourt and H. R. Delcourt, 1981, Vegetation maps for eastern North America; 40,000 yr B.P. to the present. In *Geobotany II*, ed. R. C. Romans, 123–65 (New York: Plenum Press).

5. B. Kurtén and E. Anderson, 1980, *Pleistocene mammals of North America* (New York: Columbia University Press).

6. C. R. Harington, 1978, Quaternary vertebrate faunas of Canada and Alaska and their suggested chronological sequence. Canadian National Museum of Natural Sciences, *Syllogeus 15*. Also, J. E. King and J. J. Saunders, 1984, Environmental insularity and the extinction of the American mastodont. In *Quaternary Extinctions*, ed. P. S. Martin and R. J. Klein, 315–39 (Tucson: University of Arizona Press).

7. A. Dreimanis, 1967, Mastodons, their geologic age and extinction in Ontario, Canada, *Canadian Journal of Earth Sciences* 4:663–75.

8. Kurtén and Anderson, *Pleistocene mammals*.

9. C. R. Harington and A. C. Ashworth, 1986, A mammoth (*Mammuthus primigenius*) tooth from late Wisconsin deposits near Embden, North Dakota, and comments on the distribution of woolly mammoths south of the Wisconsin ice sheets, *Canadian Journal of Earth Sciences* 23:909–18.

10. F. H. West, 1983, The antiquity of man. In *Late Quaternary environ-*

ments of the United States, vol. 2, The Late Pleistocene, ed. S. C. Porter, 364–82 (Minneapolis: University of Minnesota Press).

11. Ibid.

12. A. McS. Stalker, 1977, Indications of Wisconsin and earlier man from the southwest Canadian Prairies, *Annals of the New York Academy of Sciences* 288:119–36.

13. R. M. Brown, et al. 1983, Accelerator ¹⁴C dating of the Taber Child, *Canadian Journal of Archaeology* 7:233–7.

14. M. C. Wilson, D. W. Harvey, and R. G. Forbis, 1983, Geoarchaeological investigations of the age and context of the Stalker (Taber Child) Site, D1Pa 4, Alberta, *Canadian Journal of Archaeology* 7:179–207.

15. A. McS. Stalker, 1983, A detailed stratigraphy of the Woodpecker Island section and commentary on the Taber Child bones, *Canadian Journal of Archaeology* 7:209–22.

16. P. Morisset, 1971, Endemism in the vascular plants of the Gulf of St. Lawrence region, *Naturaliste canadien* 98:167–77.

17. N. G. Miller and G. G. Thompson, 1979, Boreal and western North American plants in the Late Pleistocene of Vermont, *Journal of the Arnold Arboretum* 60:167–218.

18. Morisset, Endemism in vascular plants.

19. The range maps are adapted from those in E. Hultén, 1968, *Flora of Alaska and neighboring territories* (Stanford, Calif.: Stanford University Press).

20. R. M. Pyle, 1981, *The Audubon Society field guide to North American butterflies* (New York: Alfred A. Knopf).

Chapter 6

1. J. V. Matthews, Jr., 1982, East Beringia during late Wisconsin time: a review of the biotic evidence. In *Paleoecology of Beringia,* ed. D. M. Hopkins, J. V. Matthews, Jr., C. E. Schweger, and S. B. Young, 127–50 (New York: Academic Press).

2. R. M. Mengel, 1970, The North American Central Plains as an isolating agent in bird speciation. In *Pleistocene and Recent environments of the central Great Plains,* ed. W. Dort, Jr., and J. K. Jones, Jr., 279–340 (Lawrence, Kan.: University of Kansas Press).

3. C. C. Lindsey and J. D. McPhail, 1986, Zoogeography of fishes of the Yukon and Mackenzie basins. In *The zoogeography of North American freshwater fishes,* ed. C. H. Hocutt and E. O. Wiley, 639–74 (New York: Wiley).

4. R. D. Guthrie, 1968, Paleoecology of the large-mammal community in interior Alaska during the late Pleistocene, *American Midland Naturalist* 79:346–63.

5. C. R. Harington, 1978, Quaternary vertebrate faunas of Canada and Alaska and their suggested chronological sequence, Canadian National Museum of Natural Sciences, *Syllogeus 15.*

6. J. Baldauf, 1982, Identification of the Holocene-Pleistocene boundary in the Bering Sea by diatoms, *Boreas* 11:113–8.

7. P. A. Colinvaux, 1981, Historical ecology in Beringia: the south land bridge coast at St. Paul Island, *Quaternary Research* 16:18–36.

8. K. R. Fladmark, 1979, Routes: Alternate migration corridors for early man in North America, *American Antiquity* 44:55–69.

9. J. J. Clague, 1981, Late Quaternary geology and geochronology of British Columbia, Geological Survey of Canada Paper 80–35.

10. J. J. Clague, J. R. Harper, R. J. Hebda, and D. E. Howes, 1982, Late Quaternary sea level and crustal movements, coastal British Columbia, *Canadian Journal of Earth Sciences* 19:597–618.

11. B. G. Warner, R. W. Mathewes, and J. J. Clague, 1982, Ice-free conditions on the Queen Charlotte Islands, British Columbia, at the height of late Wisconsin glaciation, *Science* 218:675–7.

12. C. J. Heusser, 1972, Palynology and phytogeographical significance of a Late-Pleistocene refugium near Kalaloch, Washington, *Quaternary Research* 2:189–201.

13. J. E. Muller, 1977, Geology of Vancouver Island, Geological Survey of Canada, Ottawa, Open File 463.

14. R. T. Ogilvie and A. Ceska, 1984, Alpine plants of phytogeographic interest on northwestern Vancouver Island, *Canadian Journal of Botany* 62:2356–62.

15. J. A. Calder and R. L. Taylor, 1968, *Flora of the Queen Charlotte Islands, Part 1, Systematics of the vascular plants*, Research Branch, Canada Department of Agriculture Monograph No. 4, Ottawa.

16. J. B. Foster, 1965, The evolution of the mammals of the Queen Charlotte Islands, British Columbia, British Columbia Provincial Museum Occasional Paper No. 14.

17. Fladmark, Routes.

18. A. S. Dyke and V. K. Prest, 1987, Late Wisconsinan and Holocene history of the Laurentide ice sheet, *Géographie physique et Quaternaire* 41:237–63. This paper is accompanied by a set of eleven maps, on three sheets, showing what the paleogeography of North America is believed to have been like at the following times: 18k, 14k, 13k, 12k, 11k, 10k, 9k, 8.4k, 8k, 7k, and 5k B.P. The map reference is A. S. Dyke and V. K. Prest, 1987, Paleogeography of northern North America, 18 000–5 000 years ago, Geological Survey of Canada, Map 1703A, scale 1:12 500 000.

19. F. C. Whitmore, Jr., H. B. S. Cooke, and D. J. P. Swift, 1967, Elephant teeth from the Atlantic continental shelf, *Science* 156:1477–81.

20. J. Terasmae and R. J. Mott, 1971, Postglacial history and palynology of Sable Island, Nova Scotia, *Geoscience and Man* 3:17–28.

21. H. F. Howden, et al., 1970, Fauna of Sable Island and its zoogeographic affinities: a compendium, Ottawa: National Museum of Natural Sciences, Publications in Zoology No. 4.

22. P. Morisset, 1971, Endemism in the vascular plants of the Gulf of St. Lawrence region, *Naturaliste canadien* 98:167–77.

23. G. Vilks and P. J. Mudie, 1978, Early deglaciation of the Labrador Shelf, *Science* 202:1181–3.

24. D. Dodds, 1983, Terrestrial mammals. In *Biogeography and ecology of the island of Newfoundland*, G. R. South, chapter 13 (The Hague: Dr. W. Junk Publishers).

Chapter 7

1. D. M. Hopkins, 1982, Aspects of the paleogeography of Beringia during the Late Pleistocene. In *Paleoecology of Beringia*, ed. D. M. Hopkins, J. V. Matthews, Jr., C. E. Schweger, and S. B. Young, 3–28 (New York: Academic Press).

2. C. R. Harington, 1978, Quaternary vertebrate faunas of Canada and Alaska and their suggested chronological sequence, *Syllogeus No. 15*, National Museums of Canada, Ottawa. Also, R. D. Guthrie, 1966, The extinct wapiti of Alaska and Yukon Territory, *Canadian Journal of Zoology* 44:47–57, and R. D. Guthrie, 1982, Mammals of the mammoth steppe as paleoenvironmental indicators. In Hopkins, et al., *Paleoecology of Beringia*, 307–26.

3. R. E. Redmann, 1982, Production and diversity in contemporary grasslands. In Hopkins, et al., *Paleoecology of Beringia*, 223–39.

4. D. M. Hopkins, P. A. Smith, and J. V. Matthews, Jr., 1981, Dated wood from Alaska and the Yukon: Implications for forest refugia in Beringia, *Quaternary Research* 15:217–49.

5. D. F. Murray, 1980, Balsam poplar in arctic Alaska, *Canadian Journal of Anthropology* 1:29–32.

6. Guthrie, Mammals of the mammoth steppe.

7. S. B. Young, 1982, The vegetation of land-bridge Beringia. In Hopkins, et al., *Paleoecology of Beringia*, 179–91.

8. J. V. Matthews, Jr., 1982, East Beringia during late Wisconsin time: a review of the biotic evidence. In Hopkins, et al., *Paleoecology of Beringia*, 127–50.

9. L. C. Cwynar, 1982, A Late-Quaternary vegetation history from Hanging Lake, northern Yukon, *Ecological Monographs* 52:1–24.

10. L. C. Cwynar and J. C. Ritchie, 1980, Arctic steppe-tundra: a Yukon perspective, *Science* 208:1375–7.

11. R. E. Morlan and J. Cinq-Mars, 1982, Ancient Beringians: Human occupation in the late Pleistocene of Alaska and the Yukon Territory. In Hopkins, et al., *Paleoecology of Beringia*, 353–81.

12. Harington, Quaternary vertebrate faunas.

13. E. J. Crossman and C. R. Harington, 1970, Pleistocene pike, *Esox lucius*, and *Esox* sp, from the Yukon Territory and Ontario, *Canadian Journal of Earth Science* 7:1130–8.

14. J. D. McPhail and C. C. Lindsey, 1970, Freshwater fishes of northwestern Canada and Alaska, Fisheries Research Board of Canada, Ottawa, Bulletin 173.

15. B. F. Beebe, 1978, Northern Yukon research program: Vertebrate paleontology. In *Abstracts of Fifth Biennial Meeting of AMQUA*, University of Alberta, Edmonton, September 2–4, 1978, p. 159.

16. N. W. Rutter, 1984, Pleistocene history of the western Canadian ice-

free corridor. In *Quaternary Stratigraphy of Canada—a Canadian contribution to IGCP Project 24*, ed. R. J. Fulton, 50–56, Geological Survey of Canada Paper 84–10.

17. A. MacS. Stalker, 1978, The geology of the ice-free corridor: the southrn half. In *Abstracts of Fifth Biennial Meeting of AMQUA*, University of Alberta, Edmonton, September 2–4, 1978, 19–22.

18. K. R. Fladmark, 1979, Routes: Alternate migration corridors for early man in North America, *American Antiquity* 44:55–69.

19. R. D. Guthrie, 1970, Bison evolution and zoogeography in North America during the Pleistocene, *Quarterly Review of Biology* 45:1–15. Also, R. D. Guthrie, 1980, Bison and man in North America, *Canadian Journal of Anthropology* 1:55–73.

20. J. G. Packer and D. H. Vitt, 1974, Mountain Park: a plant refugium in the Canadian Rocky Mountains, *Canadian Journal of Botany* 52:1393–1409.

21. C. B. Beaty, 1975, *The landscapes of southern Alberta* (Lethbridge, Alberta: Department of Geography, University of Lethbridge).

22. J. A. Burns, 1980, The brown lemming, *Lemmus sibiricus* (Rodentia, Arvicolidae), in the late Pleistocene of Alberta and its postglacial dispersal, *Canadian Journal of Zoology* 58:1507–11.

23. J. R. Holsinger, J. S. Mort, and A. D. Recklies, 1983, The subterranean crustacean fauna of Castleguard Cave, Columbia Icefields, Alberta, Canada and its zoogeographic significance, *Arctic and Alpine Research* 15:543–9.

24. M. Gascoyne et al., 1983, The antiquity of Castleguard Cave, Columbia Icefields, Alberta, Canada, *Arctic and Alpine Research* 15:463–70.

25. H. F. Clifford and G. Bergstrom, 1976, The blind aquatic isopod *Salmasellus* from a cave spring of the Rocky Mountains' eastern slopes, with comments on a Wisconsin refugium, *Canadian Journal of Zoology* 54:2028–32.

Chapter 8

1. H. E. Wright, Jr., 1984, Sensitivity and response time of natural systems to climatic change in the late Quaternary, *Quaternary Science Reviews* 3:91–131.

2. G. H. Denton and T. J. Hughes, eds., 1980, *The last great ice sheets* (New York: Wiley).

3. A. C. Ashworth, D. P. Schwert, W. A. Watts, and H. E. Wright, Jr., 1981, Plant and insect fossils at Norwood in south-central Minnesota: a record of late glacial succession, *Quaternary Research* 16:66–79.

4. S. Lichti-Fedorovich, 1970, The pollen stratigraphy of a dated section of Late Pleistocene lake sediment from central Alberta, *Canadian Journal of Earth Sciences* 7:938–45.

5. W. A. Watts, 1983, Vegetational history of the eastern United States. In *Late Quaternary environments of the United States, vol. 1, the Late Pleistocene*, ed. S. C. Porter, 294–310 (Minneapolis: University of Minnesota Press).

6. A. V. Morgan and A. Morgan, 1980, Faunal assemblages and distributional shifts of Coleoptera during the late Pleistocene in Canada and the

northern United States, *Canadian Entomologist* 112:1105–28. Also, D. P. Schwert, T. W. Anderson, A. Morgan, et al., 1985, Changes in late Quaternary vegetation and insect communities in southwestern Ontario. *Quaternary Research* 23:205–26.

7. C. R. Harington and A. C. Ashworth, 1986, A mammoth (*Mammuthus primigenius*) tooth from late Wisconsin deposits near Embden, North Dakota, and comments on the distribution of woolly mammoths south of the Wisconsin ice sheets, *Canadian Journal of Earth Sciences* 23:909–18.

8. T. L. Péwé, 1983, The periglacial environment in North America during Wisconsin time. In *Late Quaternary environments of the United States, vol. 1, The Late Pleistocene*, ed. S. C. Porter, 157–89 (Minneapolis: University of Minnesota Press).

9. B. Kurtén and E. Anderson, 1980, *Pleistocene mammals of North America*, (New York: Columbia University Press).

10. A. C. Ashworth and A. M. Cvancara, 1983, Paleoecology of the southern part of the Lake Agassiz basin. In *Glacial Lake Agassiz*, ed. J. T. Teller and L. Clayton, 138–56, The Geological Association of Canada Paper 26, University of Toronto Press.

11. M. J. Burke, L. V. Gusta, H. A. Quamme, et al., 1976, Freezing and injury in plants, *Annual Review of Plant Physiology* 27:507–28.

12. E. C. Pielou, 1988, *The world of northern evergreens*, (Ithaca, N.Y.: Cornell University Press).

13. Wright, Jr., Sensitivity and response time.

14. Ashworth and Cvancara, Paleoecology.

15. A. S. Dyke and V. K. Prest, 1987, Late Wisconsin and Holocene history of the Laurentide ice sheet, *Géographie physique et Quaternaire* 41:237–63.

16. A. Post and G. Streveler, 1976, The tilted forest: Glaciological-geologic implications of vegetated neoglacial ice at Lituya Bat, Alaska, *Quaternary Research* 6:111–7.

17. L. Clayton, 1967, Stagnant-glacier features of the Missouri Coteau in North Dakota. In *Glacial geology of the Missouri Coteau and adjacent areas*, ed. L. Clayton and T. F. Freers, 25–46, North Dakota Geological Survey, Miscellaneous Series 30, Grand Forks, North Dakota.

18. M-B. Florin and H. E. Wright, Jr. 1969, Diatom evidence for the persistence of stagnant ice in Minnesota, *Geological Society of America Bulletin* 80:695–704.

19. S. J. Tuthill, L. Clayton, and W. M. Laird, 1964, A comparison of a fossil Pleistocene molluscan fauna from North Dakota with a Recent molluscan fauna from Minnesota, *American Midland Naturalist* 71:344–62.

20. W. A. Watts and R. C. Bright, 1968, Pollen, seed and mollusk analysis of a sediment core from Pickerel Lake, northeastern South Dakota, *Geological Society of America Bulletin* 79:855–76.

21. A. C. Ashworth and A. M. Cvancara, 1983, Paleoecology of the southern part of the Lake Agassiz basin. In *Glacial Lake Agassiz*, ed. J. T. Teller and L. Clayton, 133–56, Geological Association of Canada Special Paper 26. Also, Tuthill, Clayton, and Laird, A comparison of a fossil.

22. Clayton, Stagnant-glacier features.

23. K. W. Stewart and C. C. Lindsey, 1983, Postglacial dispersal of lower vertebrates in the Lake Agassiz region. In *Glacial Lake Agassiz*, ed. J. T. Teller and L. Clayton, 391–419, Geological Association of Canada Special Paper 26.

24. Tuthill, Clayton, and Laird, A comparison of a fossil.

Chapter 9

1. B. F. Atwater, 1987, Status of Glacial Lake Columbia during the last floods from Glacial Lake Missoula, *Quaternary Research* 27:182–201.

2. A. S. Dyke and V. K. Prest, 1987, Paleogeography of northern North America, 18 000–5 000 years ago, Geological Survey of Canada, Map 1703A, scale 1:12 500 000.

3. R. B. Waitt, 1984, Periodic jokulhlaups from Pleistocene Lake Missoula—new evidence from varved sediment in northern Idaho and Washington, *Quaternary Research* 22:46–58.

4. J. D. McPhail and C. C. Lindsey, 1986, Zoogeography of the freshwater fishes of Cascadia (the Columbia system and rivers north to the Stikine). In *The zoogeography of North American freshwater fishes*, ed. C. H. Hocutt and E. O. Wiley, 615–37 (New York: Wiley-Interscience).

5. R. N. Mack, V. M. Bryant, Jr., and R. Fryxell, 1976, Pollen sequence from the Columbia Basin, Washington: Reappraisal of postglacial vegetation, *American Midland Naturalist* 95:390–7.

6. M. R. Kaatz, 1959, Patterned ground in central Washington: a preliminary report, *Northwest Science* 33:145–56.

7. R. A. Bodaly and C. C. Lindsey, 1977, Pleistocene watershed exchanges and the fish fauna of the Peel River basin. Yukon Territory, *Journal of the Fisheries Research Board of Canada* 34:388–95. Also, C. C. Lindsey and J. D. McPhail, 1986, Zoogeography of fishes of the Yukon and Mackenzie basins. In *the zoogeography of North American freshwater fishes*, ed. C. H. Hocutt and E. O. Wiley, 639–74 (New York: Wiley-Interscience).

8. C. C. Lindsey and J. D. McPhail, 1986, Zoogeography of fishes of the Yukon and Mackenzie basins, in *The zoogeography of North American freshwater fishes*, ed. C. H. Hocutt and E. O. Wiley, 639–74 (New York: Wiley-Interscience).

9. Ibid.

10. J. T. Teller and L. Clayton, 1983, An introduction to Glacial Lake Agassiz. In *Glacial Lake Agassiz*, ed. J. T. Teller and L. Clayton, 2–5, Geological Association of Canada Special Paper 26.

11. A. S. Dyke and V. K. Prest, 1987, Late Wisconsin and Holocene history of the Laurentide ice sheet, *Géographie physique et Quaternaire* 41:237–63.

12. Lindsey and McPhail, Zoogeography of fishes.

13. L. Clayton, 1983, Chronology of Lake Agassiz drainage to Lake Superior. In *Glacial Lake Agassiz*, ed. J. T. Teller and L. Clayton, 291–307, Geological Association of Canada Special Paper 26.

14. K. W. Stewart and C. C. Lindsey, 1983, Postglacial dispersal of lower vertebrates in the Lake Agassiz region. In *Glacial Lake Agassiz*, ed. J. T. Teller and L. Clayton, 391–419, Geological Association of Canada Special Paper 26.

15. A. E. Kehew and L. Clayton, 1983, Late Wisconsin floods and development of the Souris-Pembina spillway system in Saskatchewan, North Dakota, and Manitoba. In *Glacial Lake Agassiz*, ed. J. T. Teller and L. Clayton, 187–209, Geological Association of Canada, Special Paper 26.

16. J. T. Teller and L. H. Thorleifson, 1983, The Lake Agassiz–Lake Superior connection. In *Glacial Lake Agassiz*, ed. J. T. Teller and L. Clayton, 261–90, Geological Association of Canada Special Paper 26.

17. A. C. Ashworth and A. M. Cvancara, 1983, Paleoecology of the southern part of the Lake Agassiz basin. In *Glacial Lake Agassiz*, ed. J. T. Teller and L. Clayton, 133–56, Geological Association of Canada, Special Paper 26.

18. C. T. Shay, 1969, Vegetation history of the southern Lake Agassiz basin during the past 12,000 years. In *Life, land and water*, ed. W. J. Mayer-Oakes, 231–52 (Winnipeg: University of Manitoba Press).

19. P. F. Karrow, A. H. Clarke, and H. B. Herrington, 1972, Pleistocene molluscs from Lake Iroquois deposits in Ontario, *Canadian Journal of Earth Sciences* 9:589–95. Also, P. F. Karrow, T. W. Anderson, A. H. Clarke, et al., 1975, Stratigraphy, paleontology, and age of Lake Algonquin sediments in southwestern Ontario, Canada, *Quaternary Research* 5:49–87.

20. D. M. Lehmkuhl, 1980, Temporal and spatial changes in the Canadian insect fauna: patterns and explanations. The Prairies. *Canadian Entomologist* 112:1145–59.

21. R. M. Bailey and G. R. Smith, 1981, Origin and geography of the fish fauna of the Laurentian Great Lakes basin, *Canadian Journal of Fisheries and Aquatic Sciences* 38:1539–61.

22. J. S. Vincent and L. Hardy, 1979, The evolution of glacial Lakes Barlow and Ojibway, Quebec and Ontario, Geological Survey of Canada Bulletin 316.

23. E. J. Crossman and D. E. McAllister, 1986, Zoogeography of freshwater fishes of the Hudson Bay drainage. Ungava Bay and the Arctic archipelago. In *The zoogeography of North American freshwater fishes*, ed. C. H. Hocutt and E. O. Wiley, 53–104, (New York: Wiley-Interscience).

24. Dyke and Prest, Late Wisconsin and Holocene history.

Chapter 10

1. V. K. Prest, D. R. Grant, and V. N. Rampton, compilers, 1967, Map 1253A, Glacial Map of Canada, scale 1:5 000 000, Geological Survey of Canada.

2. D. M. Hopkins, 1982, Aspects of the paleogeography of Beringia during the late Pleistocene. In *Paleoecology of Beringia*, ed. D. M. Hopkins, R. V. Matthews, Jr., C. E. Schweger, and S. B. Young, 3–28 (New York: Academic Press).

3. T. A. Ager, 1982, Vegetational history of western Alaska during the Wisconsin glacial interval and the Holocene. In *Paleoecology of Beringia*, ed. D. M. Hopkins, R. V. Matthews, Jr., C. E. Schweger, and S. B. Young, 75–93 (New York: Academic Press).

4. S. B. Young, 1982, The vegetation of land-bridge Beringia. In *Paleoecology of Beringia*, ed. D. M. Hopkins, R. V. Matthews, Jr., C. E. Schweger, and S. B. Young, 179–94 (New York: Academic Press).

5. Hopkins, Aspects of paleogeography.

6. W. B. Workman, 1980, Holocene peopling of the New World: Implications of the arctic and subarctic area, *Canadian Journal of Anthropology* 1:129–39.

7. C. R. Harington, 1978, Quaternary vertebrate faunas of Canada and Alaska and their suggested chronological sequence, Canadian National Museum of Natural Sciences, *Syllogeus 15*.

8. M. Stuiver and H. W. Borns, Jr., 1975, Late Quaternary marine invasion in Maine: Its chronology and associated crustal movement, *Geological Society of America Bulletin* 86:99–104.

9. G. A. Bartlett and L. Molinsky, 1972, Foraminifera and the Holocene history of the Gulf of St Lawrence, *Canadian Journal of Earth Sciences* 9:1204–15.

10. C. R. Harington, 1985, Quaternary marine mammals in Canada, Abstracts of Canadian Quaternary Association (CANQUA) Biennial Conference, Lethbridge, Alberta.

11. R. B. Davis and G. L. Jacobson, Jr., 1985, Late glacial and early Holocene landscapes in northern New England and adjacent areas of Canada, *Quaternary Research* 23:341–68.

12. A. E. Roland and E. C. Smith, 1969, *The flora of Nova Scotia*, Halifax, Nova Scotia: The Nova Scotia Museum.

13. Davis and Jacobson, Jr., Late glacial and early Holocene landscapes. Also, R. J. Mott, 1975, Palynological studies of lake sediment profiles from southwestern New Brunswick, *Canadian Journal of Earth Sciences* 12:273–88.

14. D. G. Green, 1986, Pollen evidence for the postglacial origins of Nova Scotia's forests, *Canadian Journal of Botany* 65:1163–79.

15. A. S. Dyke and V. K. Prest, 1987, Late Wisconsinan and Holocene history of the Laurentide Ice Sheet, *Géographie physique et Quaternaire* 41:237–63.

16. Harington, Quaternary marine animals.

17. T. M. Cavender, 1986, Review of the fossil history of North American freshwater fishes. In *The zoogeography of North American freshwater fishes*, ed. C. H. Hocutt and E. O. Wiley, 699–724 (New York: Wiley-Interscience).

18. T. M. Cronin, 1977, Late-Wisconsin marine environments of the Champlain valley (New York, Quebec), *Quaternary Research* 7:238–53.

19. Harington, Quaternary vertebrate faunas.

20. E. J. Cushing, 1965, Problems in the Quaternary phytogeography of the Great Lakes region. In *The Quaternary of the United States*, ed. H. E. Wright, Jr., and D. G. Frey, 403–16 (Princeton, N.J.: Princeton University Press).

21. J. A. Elson, 1969, Late Quaternary marine submergence of Quebec, *Révue Géographique Montréal* 23:247–58.

22. P. W. Webber, J. W. Richardson, and J. T. Andrews, 1970, Postglacial uplift and substrate age at Cape Henrietta Maria, southeastern Hudson Bay, Canada, *Canadian Journal of Earth Sciences* 7:317–25.

Chapter 11

1. A. Neftel, J. Schwander, B. Stauffer, and R. Zumbrunn, 1982, Ice core sample measurements give atmospheric CO_2 content during the past 40,000 yr, *Nature* 295:220–3.

2. Ibid.

3. W. S. Broecker, M. Ewing, and B. K. Heezen, 1960, Evidence for an abrupt change in climate close to 11,000 years ago, *American Journal of Science* 258:429–448.

4. B. Kurtén and E. Anderson, 1980, *Pleistocene mammals of North America* (New York: Columbia University Press).

5. D. J. Meltzer and J. I. Mead, 1983, The timing of late Pleistocene mammalian extinctions in North America, *Quaternary Research* 19:130–5.

6. The map in figure 11.1 uses data from the following sources: J. G. Ogden, III, 1967, Radiocarbon and pollen evidence for a sudden change in climate in the Great Lakes region approximately 10,000 years ago. In *Quaternary Paleoecology*, ed. E. J. Cushing and H. E. Wright, Jr., 117–27 (New Haven, Conn.: Yale University Press); D. C. Amundson and H. E. Wright, Jr., 1979, Forest changes in Minnesota at the end of the Pleistocene, *Ecological Monographs* 1979:1–16; T. Webb, III, E. J. Cushing, and H. E. Wright, Jr., 1983, Holocene changes in the vegetation of the Midwest. In *Late-Quaternary environments of the United States, vol. 2, The Holocene*, ed. H. E. Wright, Jr., 142–65 (Minneapolis: University of Minnesota Press); J. C. Ritchie, 1987, *Postglacial vegetation of Canada* (New York: Cambridge University Press).

7. W. A. Watts, 1983, Vegetational history of the eastern United States. In *Late Quaternary environments of the United States, vol. 1, the Late Pleistocene*, ed. S. C. Porter, 294–310 (Minneapolis: University of Minnesota Press).

8. T. Webb, III, 1981, The past 11,000 years of vegetational change in eastern North America, *BioScience* 31:501–6.

9. M. B. Davis, 1983, Holocene vegetational history of the eastern United States. In *Late Quaternary environments of the United States, vol. 2, The Holocene*, ed. H. E. Wright, Jr., 166–81 (Minneapolis: University of Minnesota Press).

10. R. W. Mathewes and L. E. Heusser, 1981, A 12 000 year palynological record of temperature and precipitation trends in southwestern British Columbia, *Canadian Journal of Botany* 59:707–10.

11. Glacial Map of Canada, Map 1253A, scale 1:5,000,000, Geological Survey of Canada.

12. J. Terasmae, 1977, Postglacial history of Canadian Muskeg. In *Muskeg and the northern environment in Canada*, ed. N. W. Radforth and C. O. Brawner, 9–30 (Toronto: University of Toronto Press).

13. W. A. Watts and R. C. Bright, 1968, Pollen, seed and mollusk analysis of a sediment core from Pickerel Lake, northeastern South Dakota, *Geological Society of America Bulletin* 79:1339–60.

14. D. Löve, 1959, The postglacial development of the flora of Manitoba: a discussion, *Canadian Journal of Botany* 37:547–85.

15. H. E. Wright, Jr., 1970, Vegetational history of the Great Plains. In *Pleistocene and Recent environments of the central Great Plains*, ed. W. Dort, Jr., and J. K. Jones, Jr., 158–72 (Lawrence, Kan.: University of Kansas Press).

16. J. C. Ritchie, 1976. The late-Quaternary vegetational history of the western interior of Canada, *Canadian Journal of Botany* 54:1793–1818.

17. C. R. Harington, 1978, Quaternary vertebrate faunas of Canada and Alaska and their suggested chronological sequence, Canadian National Museum of Natural Sciences, *Syllogeus 15*.

18. C. S. Churcher, 1968, Pleistocene ungulates from the Bow River gravels at Cochrane, Alberta, *Canadian Journal of Earth Sciences* 5:1467–88.

19. R. S. Hoffmann and J. K. Jones, Jr., 1970, Influence of late-glacial and post-glacial events on the distribution of recent mammals on the northern Great Plains. In *Pleistocene and Recent environments of the central Great Plains*, ed. W. Dort, Jr., and J. K. Jones, Jr., 355–94 (Lawrence, Kan.: University of Kansas Press).

20. A. W. F. Banfield, 1974, *The mammals of Canada* (Toronto: University of Toronto Press).

21. J. C. Driver, 1988, Late Pleistocene and Holocene vertebrates and paleoenvironments from Charlie Lake Cave, northeast British Columbia, *Canadian Journal of Earth Sciences* 25:1545–53.

22. K. R. Fladmark, J. C. Driver, and D. Alexander, 1988, The Paleo-Indian component at Charlie Lake Cave (HbRf39), British Columbia, *American Antiquity* 53:371–84.

23. E. H. Moss (1983), *Flora of Alberta*, 2d ed., rev. J. G. Packer (Toronto: University of Toronto Press).

24. E. Hultén, 1968, *Flora of Alaska*, (Palo Alto, Calif.: Stanford University Press).

25. N. W. Rutter, V. Geist, and D. M. Shackleton, 1972, A bighorn sheep skull 9280 years old from British Columbia, *Journal of Mammalogy* 53:641–4.

26. Churcher, Pleistocene ungulates.

27. C. W. Barnosky, 1985, Late Quaternary vegetation near Battle Ground Lake, southern Puget Trough, Washington, *Geological Society of America Bulletin* 96:263–71.

28. S. S. Porter and R. J. Carson, III, 1971, Problem of interpreting radiocarbon dates from dead-ice terrain, with an example from the Puget Lowland of Washington, *Quaternary Research* 1:410–4.

29. R. W. Mathewes, 1985, Paleobotanical evidence for climatic change in southern British Columbia during Late-glacial and Holocene time, *Syllogeus* 55:397–422. Also, Barnosky, Late Quaternary vegetation.

30. C. R. Harington, 1978, Quaternary vertebrate faunas of Canada and

Alaska and their suggested chronological sequence, Canadian National Museum of Natural Sciences, *Syllogeus 15.*

31. J. D. McPhail and C. C. Lindsey, 1986, Zoogeography of the freshwater fishes of Cascadia (the Columbia system and rivers north to the Stikine). In *The zoogeography of North American freshwater fishes,* ed. C. H. Hocutt and E. O. Wiley, 615–37 (New York: Wiley-Interscience).

32. Ibid.

33. T. A. Ager, 1983, Holocene vegetational history of Alaska. In *Late Quaternary environments of the United States, vol. 2, The Holocene,* ed. H. E. Wright, Jr., 128–41 (Minneapolis: University of Minnesota Press).

34. S. B. Young, 1982, The vegetation of land-bridge Beringia. In *Paleoecology of Beringia,* ed. D. M. Hopkins et al., 179–94 (New York: Academic Press).

35. A. E. Porsild, C. R. Harington, and G. A. Mulligan, 1967, *Lupinus arcticus* Wats. grown from seeds of Pleistocene age, *Science* 153:113–4.

Chapter 12

1. B. Kurtén and E. Anderson, 1980, *Pleistocene mammals of North America* (New York: Columbia University Press). Also S. D. Webb, 1984, Ten million years of mammal extinctions in North America. In *Quaternary extinctions,* ed. P. S. Martin and R. G. Klein, 189–210 (Tucson: The University of Arizona Press).

2. P. S. Martin, 1986, Refuting late Pleistocene extinction models. In *Dynamics of extinction,* ed. D. K. Elliott, 107–130 (New York: Wiley-Interscience).

3. Webb, Ten million years.

4. Ibid.

5. Ibid.

6. D. R. Horton, 1984, Red kangaroos: Last of the Australian megafauna. In *Quaternary extinctions,* ed. P. S. Martin and R. G. Klein, 639–80 (Tucson: The University of Arizona Press).

7. E. J. Butler and F. Hoyle, 1979, On the effects of a sudden change in the albedo of the earth, *Astophysics and Space Sciences* 60:505–11.

8. Kurtén and Anderson, *Pleistocene mammals.*

9. P. S. Martin, 1984, Prehistoric overkill: the global model. In *Quaternary extinctions,* ed. P. S. Martin and R. G. Klein, 354–403 (Tucson: The University of Arizona Press). Also, Martin, Refuting.

10. F. H. West, 1983, The antiquity of man. In *Late Quaternary environments of the United States, vol. 1, The Late Pleistocene,* ed. S. C. Porter, 364–82 (Minneapolis: University of Minnesota Press).

11. Horton, Red Kangaroos.

12. Martin, Prehistoric overkill.

13. D. L. Johnson, P. Kawano, and E. Ekker, 1980, Clovis strategies of hunting mammoth (*Mammuthus columbi*), *Canadian Journal of Anthropology* 1:107–14.

14. S. L. Whittington and B. Dyke, 1984, Simulating overkill: Experi-

ments with the Mosimann and Martin model. In *Quaternary extinctions,* ed. P. S. Martin and R. G. Klein, 451–65 (Tucson: The University of Arizona Press).

15. West, Antiquity of man.

16. J. E. Mosimann and P. S. Martin, 1975, Simulating overkill by PaleoIndians, *American Scientist* 63:304–13.

17. P. S. Martin, 1982, The pattern and meaning of Holarctic mammoth extinction. In *Paleoecology of Beringia,* ed. D. M. Hopkins, J. V. Matthews, Jr., C. E. Schweger, and S. B. Young, 399–408 (New York: Academic Press).

18. D. E. Dumond, 1980, The archaeology of Alaska and the peopling of America, *Science* 209:984–91.

19. J. E. King and J. J. Saunders, 1984, Environmental insularity and the extinction of the American mastodont. In *Quaternary extinctions,* ed. P. S. Martin and R. J. Klein, 315–39 (Tucson: University of Arizona Press).

20. R. D. Guthrie, 1980, Bison and man in North America, *Canadian Journal of Anthropology* 1:55–73.

21. Kurtén and Anderson, Pleistocene mammals.

22. R. W. Graham, 1986, Plant-animal interactions and Pleistocene extinctions. In *Dynamics of extinction,* ed. D. K. Elliott, 131–54 (New York: Wiley-Interscience). Also, J. E. Guilday, 1984, Pleistocene extinction and environmental change: Case study of the Appalachians. In *Quaternary extinctions,* ed. P. S. Martin and R. J. Klein, 250–8 (Tucson: University of Arizona Press).

23. R. A. Bryson and W. M. Wendland, 1967, Radiocarbon isochrones of the retreat of the Laurentide ice sheet, Technical Report No. 35, University of Wisconsin, Department of Meteorology.

24. R. A. Kiltie, 1984, Seasonality, gestation time, and large mammal extinctions. In *Quaternary Extinctions,* ed. P. S. Martin and R. J. Klein, 299–314 (Tucson: University of Arizona Press).

25. D. W. Steadman and P. S. Martin, 1984, Extinction of birds in the late Pleistocene of North America. In *Quaternary extinctions,* ed. P. S. Martin and R. J. Klein, 466–77 (Tucson: University of Arizona Press).

Chapter 13

1. A. S. Dyke and V. K. Prest, 1987, Late Wisconsinan and Holocene history of the Laurentide Ice Sheet, *Géographie Physique et Quaternaire* 41:237–64.

2. J. D. Halfman and T. C. Johnson, 1984, Enhanced atmospheric circulation over North America during the early Holocene: Evidence from Lake Superior, *Science* 224:61–63.

3. C. R. Harington, 1978, Quaternary vertebrate faunas of Canada and Alaska and their suggested chronological sequence, *Syllogeus No. 15,* National Museum of Canada, Ottawa.

4. D. McCulloch and D. Hopkins, 1966, Evidence for an early Recent warm interval in northwestern Alaska, *Geological Society of America Bulletin* 77:1089–1108.

5. J. C. Ritchie, L. C. Cwynar, and R. W. Spear, 1983, Evidence from north-west Canada for an early Holocene Milankovitch thermal maximum, *Nature* 305:126–8.

6. D. L. Elliott-Fisk, 1983, The stability of the northern Canadian tree limit, *Annals of the Association of American Geographers* 73:560–76.

7. R. J. Hebda, 1983, Late-glacial and postglacial vegetation history at Bear Cove Bog, northeast Vancouver Island, British Columbia, *Canadian Journal of Botany* 61:3172–92.

8. S. Björck, 1985, Deglaciation chronology and revegetation in northwestern Ontario, *Canadian Journal of Earth Sciences* 22:850–71.

9. R. A. Bryson, W. M. Irving, and J. A. Larsen, 1965, Radiocarbon and soil evidence of former forests in the southern Canadian tundra, *Science* 147:46–48.

10. H. Nichols, 1976, Historical aspects of the northern Canadian treeline, *Arctic* 29:38–47.

11. J. C. Ritchie, 1983, The paleoecology of the central and northern parts of the glacial Lake Agassiz basin. In *Glacial Lake Agassiz*, ed. J. T. Teller and L. Clayton, 157–70, Geological Association of Canada Special Paper 26.

12. H. Nichols, 1969, The late Quaternary history of vegetation and climate at Porcupine Mountain and Clearwater Bog, Manitoba, *Arctic and Alpine Research* 1:155–67.

13. H. E. Wright, Jr., 1976, The dynamic nature of Holocene vegetation, *Quaternary Research* 6:581–96.

14. J. Terasmae and B. G. Craig, 1958, Discovery of fossil *Ceratophyllum demersum* L. in Northwest Territories, Canada, *Canadian Journal of Botany* 36:567–9.

15. K. W. Stewart and C. C. Lindsey, 1983, Postglacial dispersal of lower vertebrates in the Lake Agassiz region. In *Glacial Lake Agassiz*, ed. J. T. Teller and L. Clayton, 391–419, Geological Association of Canada, Special Paper 26.

16. D. P. Schwert, T. W. Anderson, A. Morgan, et al., 1985, Changes in late Quaternary vegetation and insect communities in southwestern Ontario, *Quaternary Research* 23:205–26.

17. T. G. Stewart and J.England, 1983, Holocene sea-ice variations and paleoenvironmental change, northernmost Ellesmere Island, N.W.T., Canada, *Arctic and Alpine Research* 15:1–17.

18. A. H. Clarke, Jr., D. J. Stanley, J. C. Medcof, and R. E. Drinnan, 1967, Ancient oyster and bay scallop shells from Sable Island, *Nature* 215:1146–9.

19. E. L. Bousfield and M. L. H. Thomas, 1975, Post-glacial changes in distribution of littoral marine invertebrates in the Canadian Atlantic region. In *Environmental change in the maritimes*, ed. J. G. Ogden, III, and M. J. Harvey, 47–60 (Halifax: Nova Scotia Institute of Science).

20. M. B. Davis, R. W. Spear, and L. C. K. Shane, 1980, Holocene climate of New England, *Quaternary Research* 14:240–50.

21. B. H. Luckman and M. S. Kearney, 1986, Reconstruction of Holo-

cene changes in alpine vegetation and climate in the Maligne Range, Jasper National Park, Alberta, *Quaternary Research* 26:244–61.

22. R. S. Bradley, 1985, *Quaternary Paleoclimatology* (Boston: Allen and Unwin).

23. R. S. Hoffmann and R. D. Taber, 1967, Origin and history of Holarctic tundra ecosystems, with special reference to their vertebrate faunas. In *Arctic and alpine environments*, ed. H. E. Wright, Jr., and W. H. Osburn, 143–70 (Bloomington: Indiana University Press).

24. R. S. Hoffmann and J. K. Jones, Jr., 1970, Influence of late-glacial and post-glacial events on the distribution of recent mammals on the northern Great Plains. In *Pleistocene and Recent environments of the central Great Pains*, ed. W. Dort, Jr., and J. K. Jones, Jr., 341–54 (Lawrence, Kan.: University of Kansas Press).

25. R. J. Marquis and E. G. Voss, 1981, Distributions of some western North American plants disjunct in the Great Lakes region, *Michigan Botanist* 20:53–82.

26. L. S. Thompson and J. Kuijt, 1976, Montane and subalpine plants of the Sweetgrass Hills, Montana, and their relation to early postglacial environments of the northern Great Plains, *Canadian Field Naturalist* 90:432–48.

27. E. H. Moss (1983), *Flora of Alberta*, 2d ed., revised J. G. Packer (Toronto: University of Toronto Press).

28. Thompson and Kuijt, Montane and subalpine plants.

29. K. W. Stewart and C. C. Lindsey, 1983, Postglacial dispersal of lower vertebrates in the Lake Agassiz region. In *Glacial Lake Agassiz*, ed. J. T. Teller and L. Clayton, 391–419, Geological Association of Canada Special Paper 26.

30. D. M. Lehmkuhl, 1980, Temporal and spatial changes in the Canadian insect fauna: Patterns and explanations—the prairies, *Canadian Entomologist* 112:1145–59.

31. W. B. Workman, 1980, Holocene peopling of the New World: Implications of the arctic and subarctic data, *Canadian Journal of Anthropology* 1:129–39.

32. B. Kurtén and E. Anderson, 1980, *Pleistocene mammals of North America* (New York: Columbia University Press).

33. W. W. FitzHugh and H. F. Lamb, 1986, Vegetation history and culture change in Labrador prehistory, *Arctic and Alpine Research* 17:357–70.

34. L. F. Pettipas and A. P. Buchner, 1983, Paleo-Indian prehistory of the Glacial Lake Agassiz region in southern Manitoba, 11 500 to 6 500 B.P. In *Glacial Lake Agassiz*, ed. J. T. Teller and L.Clayton, 421–51, Geological Association of Canada Special Paper 26.

35. R. D. Guthrie, 1980, Bison and man in North America, *Canadian Journal of Anthropology* 1:55–73.

36. Ibid.

37. Pettipas and Buchner, Paleo-Indian prehistory.

38. C. R. Harington, 1978, Quaternary vertebrate faunas of Canada and Alaska and their suggested chronological sequence, *Canadian National Museum of Natural Sciences*, *Syllogeus 15*.

Chapter 14

1. J. Imbrie and K. P. Imbrie, 1979, *Ice ages: Solving the mystery* (Short Hills, N.J.: Enslow Publishers).

2. G. H. Denton and W. Karlen, 1973, Holocene climatic variations—Their pattern and possible cause, *Quaternary Research* 3:155–205. Also, J. M. Ryder and B. Thomson, 1986, Neoglaciation in the southern Coast Mountains of British Columbia: Chronology prior to the late Neoglacial maximum, *Canadian Journal of Earth Sciences* 23:273–87.

3. H. P. Hansen, 1952, Postglacial forests in the Grande Prairie–Lesser Slave Lake Region of Alberta, Canada, *Ecology* 33:31–40.

4. T. Alderman, 1965, It's a nuisance, *Imperial Oil Review* 49(3):6–10.

5. M. L. Heinselman, 1970, Landscape evolution, peatland types, and the environment in the Lake Agassiz Peatlands Natural Area, Minnesota, *Ecological Monographs* 40:235–61.

6. J. Terasmae, 1977, Postglacial history of Canadian muskeg. In *Muskeg and the northern environment in Canada*, ed. N. W. Radforth and C. O. Brawner, 9–30 (Toronto: University of Toronto Press).

7. A. Banner, J. Pojar, and G. E. Rouse, 1983, Postglacial paleoecology and successional relationships of a bog woodland near Prince Rupert, British Columbia, *Canadian Journal of Forest Research* 13:938–47.

8. T. A. Ager, 1983, Holocene vegetational history of Alaska. In *Late-Quaternary environments of the United States, vol. 2, The Holocene*, ed. H. E. Wright, Jr., 128–41 (Minneapolis: University of Minnesota Press).

9. W. M. Last and J. T. Teller, 1983, Holocene climate and hydrology of the Lake Manitoba basin. In *Glacial Lake Agassiz*, ed. J. T. Teller and L. Clayton, 333–53, Geological Association of Canada Special Paper 26.

10. R. B. Brugam, 1983, Holocene paelolimnology. In *Late-Quaternary environments of the United States, vol. 2, The Holocene*, ed. H. E. Wright, Jr., 208–21 (Minneapolis: University of Minnesota Press).

11. W. M. Last and T. H. Schweyen, 1985, Late Holocene history of Waldsea Lake, Saskatchewan, Canada, *Quaternary Research* 24:219–34.

12. M. B. Davis, 1983, Quaternary history of deciduous forests of eastern North America and Europe, *Annals of the Missouri Botanical Garden* 70:550–63.

13. M. B. Davis, 1983, Holocene vegetational history of the eastern United States. In *Late-Quaternary environments of the United States, vol. 2, The Holocene*, ed. H. E. Wright, Jr., 166–81 (Minneapolis: University of Minnesota Press).

14. Davis, Holocene vegetational history. Also M. B. Davis, 1981, Outbreaks of forest pathogens in Quaternary history. In *Proceedings of the Fourth International Palynological Conference, Lucknow (1976–77)*, vol. 3, 216–27.

15. D. S. Slater, 1985, Pollen analysis of postglacial sediments from Eildun Lake, District of Mackenzie, N.W.T., Canada, *Canadian Journal of Earth Sciences* 22:663–74.

16. G. M. MacDonald, 1987, Postglacial vegetation history of the Mackenzie River basin *Quaternary Research* 28:245–62.

17. H. Nichols, 1976, Historical aspects of the northern Canadian tree-line, *Arctic* 29:38–47.

18. Ibid.

19. D. L. Elliott-Fisk, 1983, The stability of the northern Canadian tree limit, *Annals of the Association of American Geographers* 73:560–76.

20. C. J. Heusser, 1954, Nunatak flora of the Juneau Ice Field, Alaska, *Bulletin of the Torrey Botanical Club* 81:236–50.

21. A. M. Swain, 1978, Environmental changes during the past 2000 years in north-central Wisconsin: Analysis of pollen, charcoal and seeds from varved lake sediments, *Quaternary Research* 10:55–68.

22. J. C. Bernabo, 1981, Quantitative estimates of temperature changes over the last 2700 years in Michigan based on pollen data, *Quaternary Research* 15:143–59.

23. H. Nichols, 1969, The late Quaternary history of vegetation and climate at Porcupine Mountain and Clearwater Bog, Manitoba, *Arctic and Alpine Research* 1:155–67.

24. R. W. Mathewes and L. E. Heusser, 1981, A 12 000 year palynological record of temperature and precipitation trends in southwestern British Columbia, *Canadian Journal of Botany* 59:707–10.

25. R. E. Vance, D. Emerson, and T. Habgood, 1983, A mid-Holocene record of vegetative change in central Alberta, *Canadian Journal of Earth Sciences* 20:364–76.

26. J. C. Ritchie, 1984, *Past and present vegetation of the far northwest of Canada* (Toronto: University of Toronto Press).

27. Bernabo, Quantitative estimates.

28. A. H. Clarke, D. J. Stanley, J. C. Medcof, and R. E. Drinnan, 1967, Ancient oyster and bay scallop shells from Sable Island. *Nature* 215:1146–49.

29. B. H. Luckman, 1986, Reconstruction of Little Ice Age events in the Canadian Rocky Mountains, *Géographie physique et Quaternaire* 40:17–28.

30. B. M. Bergsma, J. Svoboda, and B. Freedman, 1984, Entombed plant communities released by a retreating glacier at central Ellesmere Island, Canada, *Arctic* 37:49–52.

31. Anonymous, 1985, The prehistory of the Lockport site. Historic Resource Branch, Culture, Heritage and Recreation, Winnipeg.

Index

(All organisms are indexed under their English names except for those that have only Latin names. Appendix 1 lists the English names alphabetically and gives their Latin equivalents. Appendix 2 lists the Latin names alphabetically and gives their English equivalents.)